Bournemouth
Libraries

BOURNEMOUTH LIBRARIES
610690480 Y

Farmer's Boy

MICHAEL HAWKER

Old Pond Publishing

First published 2005

ISBN 1-903366-86-0

A catalogue record for this book is available from the British Library

Published by
Old Pond Publishing
Dencora Business Centre
36 Whitehouse Road
Ipswich
IP1 5LT
United Kingdom

www.oldpond.com

Cover design by Liz Whatling
Printed and bound in Great Britain by Biddles Ltd, King's Lynn

Contents

Acknowledgements

I am indebted to those who searched their archives and collections to provide photographs for my book: Clare Fisher of Beaford Arts; Julian Vane at the Museum of Barnstaple and N. Devon; Les Franklin at the North Devon Athenaeum; Verlie Meek and Jenny Jenkins at the splendid little museum at Braunton; and to the Avery family at Georgeham for permission to include a family photo.

I owe much to the farmers, their wives and my fellow workers at my two farms. Many of them have passed away, including my good friend George who taught me so many of the practical farming skills I acquired and passed on his love of the land and farming to me. They all worked hard, often in arduous conditions and were justifiably proud of their skills and ability. They displayed a conscientious approach to their duties and concern for the welfare of the stock and the horses that often extended beyond normal working hours. I learnt much of value from them.

I was privileged to know them, to work amongst them and to have been one of their number.

Introduction

The Early Days

LITTLE did I realise on a warm sunny afternoon in the late 1940s when I wandered into the cornfield, which I found out later was known as Goosey, that it would change my whole life. I was just 12, and the long school holiday had just begun. Living at the very edge of the town with fields all around I had few playmates, and the one boy who was a regular companion was visiting his granny in a neighbouring village, as he did twice a week. I was on my own, as I often was, and bored; so I walked the few yards from my house down the road, past the thatch-roofed farm buildings at the corner and on to the gate of the field from where I could hear the sounds of harvesting.

In the field a team of workers was cutting corn with a tractor and binder. One of the workers was the farmer's son, a recently married man in his early twenties who was destined to become a friend for life. George was the older brother I never had and the one who was to teach me most of the practical farming skills that I was to acquire. He was 'stooking', that is picking up the sheaves discharged by the binder and putting six together to form the stook. I began to assist by collecting the sheaves from the edge of the field and carrying them to him. He chatted, more as a friend than as a man talking to a boy. I returned next day, and as the holiday wore on I spent more and more time at the farm and helped with many different jobs, so much so that at the end of the holidays the farmer, George's father, gave me £5, my first wage. This was a considerable sum when compared with the weekly wage of a farm worker, at that time, I believe, only £3 12*s* (£3.60) per week.

From then on I spent all my school holidays and every Saturday at the farm. I learnt many different skills, and by the time that I was in my mid-teens I was working side by side with George and the other workers as a full member of the team.

My love of the farm determined the direction my career would take and I chose to study for a degree in agriculture at Reading University. As a pre-requirement of the course at Reading, I spent 14 months at another farm as a full-time worker.

After gaining my degree, I realised that with no capital I had little hope of farming on my own, and a career as a farm bailiff or farm manager offered poor long-term prospects. My studies had also made me realise that my principal interest lay in farm mechanisation rather than general agriculture, so I decided to study agricultural engineering at postgraduate level at Newcastle. This led to an MSc and an offer to carry out a programme of research into aspects of environment control in intensive livestock housing that were causing severe problems for the industry at that time. There was the added inducement of a PhD at the end if the work and final report were at the required level. I was also given the chance to do some lecturing and practical instruction in parallel with the research. My career from then on involved research, advisory work, teaching and finally, just before I retired, estate and property management.

Chapter 1

Corn Harvesting

CORN harvesting may seem to be the end of the crop year and an illogical point to begin memories of the farming scene. For me, however, it was always the most exciting part of the farming year when all the planning, work and waiting brought their rewards. Also, of course, it was the time when I wandered into the cornfield and began the interest in farming that led to my career and a lifelong friendship.

When I first entered Goosey the scene that I witnessed had altered little in the previous 40 years. The only change to indicate it was the 1940s and not 1900 was that a tractor had replaced the horses which until the previous year had pulled the binder.

As on most farms in North Devon in the mid-1940s, corn was still being cut with a self-binder. Ours was a very ancient Deering with a 4 ft 6 in (1.5 m) cutterbar and designed to be pulled by three horses, although by the time that I appeared on the scene the horse pole had been converted to a tractor drawbar. The cutterbar was offset from the machine so that the tractor and the main body of the binder did not run on the uncut crop. The corn was cut by a 'knife' fitted with triangular blades that reciprocated along the cutterbar, which was equipped with forward-projecting pointed fingers. As the knife sections slid backwards and forwards over the edges of the fingers, the scissors-like action cut the corn stems.

The cut corn stalks fell onto a moving canvas belt that conveyed them sideways and then upwards between a further pair of canvas belts to the binding mechanism that divided the crop flow into sheaves of uniform size, tied them with a band of binder twine and then ejected them. Two large balls of binder twine were carried in a canister on the binder and the twine was fed automatically to the knotter as required. The knotter was an ingenious little mechanism that automatically formed the knot and then cut the twine. The binder twine was a coarse sisal cord that was always saved after the sheaves had been threshed as it was universally used for tying up anything and everything around the farmyard.

A wide, large-diameter wheel called the 'bull' wheel supported the weight of most of the machine and drove all the moving parts through a series of chains and gears, all of which had to be greased or oiled each day before starting work.

The farmer rode on a seat high up at the rear of the binder that gave him a good view of the operation. From there he could reach all the controls needed to make operating adjustments. The machine was pulled by a Standard Fordson tractor that was usually driven by his daughter-in-law. The tractor driver determined the rate of progress. Any changes in speed needed, or a stop because of a blockage, resulted in much shouting and gesticulating by the farmer. This was usually of little avail because the driver seldom looked round and was quite oblivious to what was going on behind her, often much to the amusement of we onlookers.

Before the binder began to cut, the field was 'opened up' by cutting a swath with a scythe all around the field next to the hedge. This enabled the tractor to make the first circuit of the field without running through standing corn and knocking grain out of the ears. Ripe grains shed very easily and care was taken at every stage of harvesting to reduce this potential loss. The scythed corn was bound into sheaves by hand, using a small bundle of stalks as a band, the ends twisted together and turned in underneath to secure it. The sheaves were stood against the hedge out of the way of the tractor on its first circuit of the field.

Scything was a skilled job. The scyther swept the long blade round in an arc, cutting a swath the full width of the tractor before shuffling forward and taking the next stroke. The skill lay in maintaining good balance and a steady rhythm, and in keeping the blade very keen by re-sharpening it frequently. It was usual to strap a supple stick cut from the hedgerow and bent into a bow to the handle. This swept the cut corn to one side on each stroke, making it easier to pick up and tie into sheaves.

Although I never used the scythe in the cornfield, I frequently used one to cut areas of weed, particularly thistles, in grass fields. Until I acquired the knack it was very tiring work. The blade was fitted to a long, curved shaft fitted with two handles that could be adjusted to suit the height of the worker. This arrangement enabled the scyther to control the movement of the blade with great accuracy. The secret was to take a measured regular swing that did not take too big a bite and avoided digging the point of the blade into the ground.

The blade was sharpened with a carborundum stone. This was about 12 in. (30 cm) long and tapered towards each end. The scythe was placed with the end of the handle on the ground so that the blade was at shoulder level. While holding the end of the blade with one hand, the workman rubbed the stone along the blade, one stroke along the topside and the next along the underside. This was a potentially dangerous operation as, while sharpening the underside, it was very easy to slide the forefinger gripping the stone along the blade. As proof of this I still have the scar on my forefinger where I sliced it to the bone when sharpening a scythe.

The farmer lovingly maintained the binder. It was greased and oiled meticulously every day before work commenced. It was the most complex item of machinery on the farm at the time but, because of its age, most of the many moving parts were worn. On the whole this seemed to make little obvious difference to its efficiency. Apart from general rattles and threshing drive chains, the worst symptom of wear was the machine's irritating tendency to produce very tiny sheaves from time to time. This happened when the trip mechanism that started the knotting cycle and determined the size of the sheaf failed to stop after tying and ejecting a sheaf. This started another knotting cycle without pausing, resulting in a baby sheaf being kicked out.

The farmer was convinced of the need to have re-sharpened knives always ready for regular replacement. One worker did nothing else but sharpen them, a sinecure we stookers thought, nearly as easy as riding on the machines. The cutting knife consisted of triangular blades with two cutting edges riveted to the knife back. The knife was clamped onto a portable trestle with two screw clamps, and both edges of every blade were sharpened in turn. We usually used a file for this; but the quality of blades varied during the war and occasionally very hard blades were encountered that could only be sharpened with a carborundum stone. It was important to retain the original cutting angle of the blades and to remove any burrs formed on the flat underside of the blade by the sharpening operation. It was only when it was pointed out to me one day when I was sharpening a knife that I learned that a file only cuts on the forward stroke. I had been vigorously pushing the file backwards and forwards as if I were using a carborundum stone. Looking back now, I realise how much I was taught and how many useful practical skills I learned on the farm.

When individual blades were damaged or seriously worn down by use and by many re-sharpenings, they had to be replaced. The two rivet heads were cut off with a cold chisel and hammer, and the rivets were punched out. New rivets were pushed through the holes in the cutterbar which was placed on an anvil or some other convenient hard surface. The new blade was fitted and held tightly in place. A sharp blow with the hammer squashed out the rivet to fill the hole and prevent the blade from moving. The blow also flattened the end to form the retaining head which was then neatly rounded over using the ball end of the hammer. Rivet 'setts' were available to simplify this task and form a perfect hemispherical head, but I never saw them used on any local farm.

Once the binder had travelled five times around the field and there were six rows of sheaves on the ground, including the one in the hedge, 'stooking', or stacking as we called it, could begin. The sheaves were picked up and stooked in sixes, or occasionally eights or tens for wheat. The objective was to stand the sheaves up and allow the wind to speed up drying so that they could be carted home as soon as possible.

Ideally, three workers would form a team. The leader walked down between the third and fourth rows picking up a sheaf from each row, one under each arm. He set them down midway between the two rows with the ears pressed together and with the butts apart to form an inverted vee. His companions collected the two rows on each side of their leader and placed their pairs of sheaves at each end of the first pair to complete the stook and achieve a tent-like shape. This construction allowed airflow through the stook to speed up the drying process. The second and third members of the team had further to walk and so the leader of the team was also the pacemaker. Once a complete circuit of the field had been made, the stookers moved over to pick up the next six rows, and so on until they reached the centre of the field and all the sheaves had been stooked.

Our small fields of four to ten acres, surrounded by high banks, topped with hedges and with trees dotted along them, were typical of North Devon, and so any stooks set up near the hedges would be sheltered from wind and sun and the sheaves would be slow to dry. The first row of stooks around the field was therefore placed between the fifth and sixth rows out from the hedge. This necessitated long walks to collect the sheaves dropped in the hedgerow when the field was opened out, and my self-appointed task on that first day when I ventured into Goosey, and on many days after, was to fetch these for the other workers.

Very rarely were we able to muster more than one three-person team, and occasionally there were only two of us stooking. Often George's younger unmarried sister was pressed into service to make up the team. We three stooked many acres during the years that I helped to harvest. Only when the crop was exceptionally thin were we able to keep up with the binder and finish stooking soon after it had finished cutting. When I worked on my second farm, after I left school, I stooked many acres on my own. Entering a field which had been cut the previous day without any stooking and seeing the hundreds of sheaves waiting to be collected and stooked was daunting. However, an occasional worker at my second farm, who had stooked on the Wiltshire downs, told me that the fields there were so large and it took them so long to complete a circuit of a field that the stookers carried their lunch bags on their backs.

Ripe corn stalks are rough and scratchy, and so we always buttoned down the sleeves of our shirts when stooking. Friends or holidaymakers who came to help us from time to time and chose to remove their shirts or to wear shorts soon had scratched and reddened arms and legs to show for it. Barley and wheat were the worst offenders. Barley *hailes*, as we called the awns that formed the beard of barley ears, were a nuisance to all. As barley was cut when fully ripe, the awns were dry and brittle and had saw-toothed edges. The teeth projected at an angle and so the wretched things worked their way everywhere, forming itchy ruffs around the tops of socks and collecting where clothing was tightest. The angled teeth behaved like barbs and were awkward to remove from clothing.

Even worse were the harvest mites that seemed to occur in every cornfield. These caused reddening lumps that itched intolerably where clothing was tight, usually around the waist under the wide belts that we all wore.

However, there were compensations for these discomforts. Working together with one or two others, chattering and bantering, particularly with a good friend and in warm sunshine, was a great life. Nonetheless, it could be very hot and humid in the small fields as the high hedges and banks prevented any refreshing breezes from reaching us. We were usually running with sweat by the time that we had completed a circuit of the field and reached the gateway where a cooling drink awaited us. There might be bottles of squash, cold tea, or for a rare treat, fizzy lemonade. Cold tea may seem a curious choice but it is remarkably refreshing. Current concerns about the effects of having too many drinks containing caffeine were not prevalent then.

There were weed problems in many of our cornfields. Many of them had been permanent pastures ploughed up at the direction of the War Agricultural Executive Committee early in the war to provide more grain for the nation's larder. This resulted in a high population of weeds, particularly perennials such as docks and thistles, or *dashels* as we called them. It also resulted in areas where the crop was very thin. The concept of chemical weed control was in its infancy, and although a few herbicides were available we did not spray. The only alternatives were hoeing, hand weeding, and traditional cultivation methods used in the autumn before sowing. As a result the sheaves often contained dashels that had became dry and brittle. The presence of these made the sheaves very painful and difficult to grasp, and the prickles made your fingers sore, particularly if the wounds festered. Sometimes we picked up the sheaves by their bands but this made it difficult to stook them properly. If the infestation was very bad, we wore gloves but the prickles often penetrated these. The gloves we had were old, and the stitching along some of the fingers was pulled out so that they provided only limited protection.

Rabbits in those pre-myxomatosis days were a problem for farmers, as they still are today. In cornfields they sometimes grazed bare the edges of a field. Once cutting began, any rabbits in the field were too frightened by the noise and activity to leave the shelter of the crop still standing. Not until only a small area still remained to be cut at the centre of the field did they break cover and bolt for the safety of the hedge. We stopped stooking when this was imminent and spaced ourselves around the area of corn still standing ready to dash after any rabbit that darted out. Surprisingly perhaps, it was possible to run down a rabbit before it reached the hedge. The sheaves in their path were too big for them to jump over and they had to zigzag around them or scramble over them, whereas we could pursue them without breaking stride. Having caught one, we despatched it quickly and humanely with a rabbit punch to the neck. Meat was still only available on ration at that time and a rabbit was a welcome addition to the larder. Unfortunately, since myxomatosis wild rabbit is not generally considered to be acceptable for the table.

Once or twice a couple of men turned up with shotguns to shoot rabbits running out of the corn. Whether they were invited or whether they found us at work by chance I never discovered, but we stookers disliked them being there. They would stand some distance into the field and at opposite corners of the shrinking rectangle of uncut corn so that they could cover all four sides and most of the field. We had to continue stooking of course, and the idea of one of them swinging round to sight on a rabbit fleeing past us towards the hedge and discharging his gun in our general direction was quite disturbing. Furthermore, we looked upon any rabbits that we could catch as our perk.

The two farms on which I worked did not have a serious rabbit problem but there were fields only a few miles distant where the cornfield tally was said to regularly exceed a hundred.

Rabbiting after lunch on Boxing Day with ferrets and nets was a tradition on our farm, although digging out a ferret that had cornered and killed a rabbit in a hole, and decided not to come out, often seemed to take up most of the afternoon. At one time, two ferrets were kept in enclosures rather like rabbit hutches in the farmyard. I had little to do with them. They were smelly, even when their hutches were cleaned regularly, and they had to be handled with great care as they would give a nasty bite given half a chance.

Pigeons were another serious pest which did enormous crop damage and shoots to reduce their numbers were organised regularly. In spring nests were blasted with a shotgun to reduce the number of young birds hatched. Pigeon pie also appeared on the menu of many countryfolk.

In its working position the binder was too wide to pass through the gateways and to travel on the road, so it had to be 'converted' once cutting was complete. We had to turn it through 90 degrees, unlatch and reposition the drawbar, and then fit two transport wheels.

The shortage of natural rubber meant that most tractors purchased during the latter part of the war were fitted with steel wheels. Ours was no exception. Each rear wheel was equipped with two rows of spade lugs bolted alternately around the circumference of the wheel to reduce wheel slip on wet and sticky soils. 'Spuds', as we called these lugs, were wedge-shaped castings several inches deep and wide that dug into the soil to give grip and improve traction. To prevent these damaging the road surface during transit, we had to bolt wooden road bands shod with iron around each wheel. This did not take long in the summer when the ground was dry but in winter, when we were ploughing or cultivating, the gaps between the spuds became filled with sticky mud. We had to gouge this out with a flat steel blade carried on the tractor before we could put the road bands on. When we had removed the mud from as much of the perimeter of the wheel as possible, we could bolt on two or sometimes three of the four road bands for each wheel. Then the tractor had to be started up and moved forward so that the rest of each wheel could be cleaned and the last bands fitted. At the end of a long winter's day in the cold and wet and in the gathering gloom this was a tedious and difficult task, often resulting in damaged and bleeding knuckles when the spanner slipped off an almost invisible muddy nut hidden behind the rim of the wheel.

Occasionally, if it was only a short distance to the next field, we risked the wrath of the local constabulary and drove to it without fitting the road bands, running one wheel on the verge to reduce the number of marks that the spuds made on the road. It was a very bumpy ride. The local policeman took a more serious view of transgressions like this and of leaving mud on the road than appears to be the case today. If we did drop mud or muck on the road when we left a field, we were expected to make a reasonable attempt to scrape it up.

The length of time corn stayed in the stook depended on the stage at which each type was harvested. Barley was cut nearly fully ripe and so it could be carted to the farmyard almost immediately. Wheat could be carted within a few days, but oats shed the grain very readily before the crop was ripe, so it was cut relatively green and allowed to ripen in the stook. An old adage was that oats should hear the church bells rung thrice before being ready to cart, that is, at least two weeks.

Ripeness was determined by rubbing out the grain from an ear and biting a grain to check the hardness. Barley ears bent over very sharply when nearing ripeness so that they faced downwards, referred to as being *goosey-neckèd*. However, in every case the period in the stook was affected by the time taken for any green weed in the sheaf to dry out and for the sheaves themselves to dry out if they were rained on while in the stook. If the sheaves became very wet, they had to be re-stooked on dry ground. If wet in the middle, the sheaves might have to be turned inside out to bring the corn at the centre to the outside. This was a desperate last resort because of the time and effort involved. I can only recall doing it once or twice. The last occasion was in a dense crop of black oats that had grown exceptionally tall. It was so tall that I could barely see over it, and the ears were the longest I have ever seen, twice as long as normal oat ears. It was a very wet harvest and we struggled with that crop for weeks.

The ears were so heavy and the straw so long that the crop had *gone lie,* that is, had been flattened by the heavy rain and wind. Cutting it was a nightmare. The crop in different parts of the field was flattened in different directions, and some areas lay in whorls. The binder had to be driven in the opposite direction to the way the crop lay, with the cutterbar right down on the ground to try to get under it and cut the stalks. Instead of cutting around the field as normal, we were forced to cut areas piecemeal, in any direction, as best we could. We tried fitting lifters to the cutterbar. These were fingers fitted to the front of the cutterbar to slide under flattened straw to lift it high enough to be cut. However, the crop was so flattened that even the lifters could not always get under it and, in desperation, we had to resort to scything these areas and tying the sheaves by hand.

Even when the binder managed to cut the corn, it could not cope with it. The crop was so heavy and the straw so long that the knotters and ejectors worked continuously. The straw was in such a tangled mass that the sheaves were not being separated and ejected individually. Each sheaf kicked out was tangled with the one following it and was dragged along the ground until someone walking beside the binder pulled them apart. Stooking was complicated by the difficulty of knowing which was the top and bottom of each sheaf, and the stooks we built looked more like heaps of dark, ragged straw than stooks. I have never to this day seen another crop of oats to equal it; black oats were never grown again at home.

I continued to harvest during the summer vacations until I left university. By that time, in the 1950s, the self-propelled combine harvester that cuts and threshes the crop in one operation had reached North Devon, although much of the local corn was still cut with the binder. This transformed the harvesting scene. The cutterbar of the combine was mounted across the front of the machine, so it was no longer necessary to scythe around the field before starting to cut and, of course, stookers were redundant. A medium-density baler was used to collect and bale the threshed straw, and so we carted bales of straw instead of sheaves of corn. Nor did we have to sharpen the cutting knife as the blades riveted to it had serrated edges and were self-sharpening.

The Massey-Harris 726 combine that was used at home was a contract machine driven by the contractor. It was a 'bagger' model. Unlike modern machines that convey the threshed grain into a large tank which is emptied into a trailer from time to time, this model delivered the grain into sacks. My first job was to stand on a platform at the rear of the machine and detach each sack when it was full and then clip an empty one over the grain spout in its place. Once I had tied the neck of the filled sack, it was pitched off the combine, if possible in line with previous ones so that they were conveniently positioned for loading onto a trailer and carting home to the farm.

It was always possible to see where the combine was working from the cloud of dust that enveloped it, with the driver and I in the midst of it. Modern combines are fitted with air-conditioned cabs but we climbed down from our machine covered in dust from head to toe.

I suppose I was lucky to escape a lung condition or an allergy. If I had spent more years in practical farming, perhaps I would not have been as lucky and have developed farmer's lung, a serious debilitating illness which many farmers and their workers contracted. In those days working and labour conditions that would not be countenanced today were simply accepted. Younger men in particular, foolishly I now realise, boasted of their ability to carry out heavy and difficult jobs in adverse conditions. There was even an element of competitiveness between workers in some situations. I can remember at one farm where I helped thresh, pitching the sheaves to the threshing machine and being determined to finish my half of the cornstack first. I didn't as I recall but, of course, the rival pitcher was older and much more experienced than I was!

Chapter 2

Harvest Home

WHEN I started in the cornfield, the sheaves were transported to the rickyard across the road from the farmyard by horse and cart. It was only several years later, when it became possible to obtain rubber-tyred wheels for the tractor, that the tractor and trailer replaced the horses.

Two types of load-carrying carts were used on the farms in North Devon, the butt and the hay cart. Both were pulled by a single horse and had a single axle at the centre so that the load was balanced over the axle and carried by the wheels, not on the back of the horse. The larger four-wheeled haywains and wagons used in other parts of the country were too awkward for the narrow gateways, sunken lanes and steep sloping fields around us. The hay carts had low sides each constructed of vertical wooden rods topped by a horizontal rail. At the front and back of the cart there was a removable, sloping, high 'laithe', or 'lade', sometimes called a ladder or gate in other counties, the term 'gate' giving the best impression of their appearance. The large-diameter wooden wheels with iron rims were set outside the cart body so that the floor resting on the axle was almost at waist height. This made it quite difficult for us to climb up into the cart, although the sides did not extend the full length of the body so that there were gaps at back and front that made it slightly easier. One common method was to use a convenient spoke of the wheel as a step, although this could be quite dangerous if the horse decided to move at the wrong moment.

The wooden hubs on all our cart wheels were loose fits on the iron axles, probably due to wear; the to-and-fro sideways movement that this caused when travelling across rough or uneven ground created a distinctive clatter. The wheels were removed regularly to grease the axles and so reduce wear. The height of the axle meant that a special jack had to be used. It consisted of a wooden pillar, mounted on a base with a wooden bar pivoted at the top that acted as a lever. The short end of the bar was positioned under the axle, and the other end was pushed down to raise the cart. A chain hooked back to the pillar held it down securely. To remove the wheel, it was turned until a slot in the hub was centred over the top of the securing pin. Once the pin was withdrawn the wheel could be slid off the axle.

The procedure for harvesting depended on the size of the team available. In the field, one or two workers pitched the sheaves up to the person making the load on the cart. When the load had been completed and roped, one of the workers would lead the horse and cart back to the farm and into the rickyard where he would climb up onto the load and pitch it off onto the 'rick', or cornstack. This arrangement worked best when the team was large enough for all three carts to be in use, one in the field loading, one in the rickyard unloading and the third on the road between the two.

At my first harvest, when I was 11, I made myself useful in the field by leading the horse from one stook to the next as the sheaves were loaded onto the cart. This was quite easy because the horses knew as well as I did what was required of them and required minimal guidance, which was just as well because I had to reach up to grasp the bridle and was rather overawed by their size.

At the end of the war our farm, like most in the area, was understaffed and it was difficult to find enough workers to operate three carts. It was almost inevitable that by the time I was in my mid-teens I had been given the responsibility of my own horse and cart, making the load or pitching in the field, leading it back to the rickyard, pitching it off and returning to the field. The last was the best part because I could climb into the cart and ride back, driving the horse rather than leading her – as it was usually Flower, our only mare, that was allocated to me.

It was preferred that you made your own load and pitched it off, so that you were forced to sort it out yourself when unloading if it wasn't made properly. However, the farmer's daughter-in-law was often brought in to help and she made all the loads, and so I pitched instead alongside the farmer. He always pitched in the field as the rate at which the carts were loaded set the work rate for the whole team.

The sheaves were pitched up two at a time with long-handled *picks,* called pitch forks or hay forks elsewhere, with either two or three curved prongs. The first sheaves were laid along the sides of the cart in a row, butts outward, still in pairs. The next two or three rows partially overlapped the previous row until the centre was reached, so that each row tied in and secured the previous one. Successive layers of sheaves were laid on top of the first in the same manner, still in pairs, until the load reached its full height which was well above the front and back lades. Once the load had risen above the low sides of the cart, the subsequent rows of sheaves were pushed out to overhang the sides to increase the width and size of the load.

The load was always roped to prevent any of it being dislodged or bounced off as the un-sprung carts crossed rough or rutted ground and lurched through gateways. Each end of a rope was permanently tied to the cart shafts to form a vee. A second rope, fastened to the back of the cart at the centre, was passed through the vee and pulled down hard and tied using a carter's hitch. Once the load was roped, the load maker could slide down onto the horse's back, clinging to the ropes for support, then onto the shaft before dropping down onto the ground.

On one occasion, whilst the farmer was pulling the rope tight, it broke and he fell on his back on the stubble. He was not a young man but he climbed to his feet and carried on, after checking that the two bottles of beer which he habitually carried in the pockets of his jacket were not broken. No matter what the weather was or how hot it was, he always wore the same clothes: breeches and black boots and leggings, a black jacket that he only occasionally removed in the field, a black waistcoat, collar and tie, and a cap on his head. He was a farmer of the old school who had started with nothing but by dint of hard work, much of it in Canada, had been able to acquire a farm which was quite substantial by the standards of the time in North Devon. He was not one to mince his words, and while I was still a boy I was slightly apprehensive of him although he never said a harsh word to me. His conversation was full of quaint old Devonshire expressions that would be considered archaic today. For instance, he still referred to young women as maidens, and might use a phrase such as 'cum yer, maid'.

It was a great loss of face to lose a load. I never did, although several years later George and I did manage to spread bales along the roadside when we lost part of a trailer load.

Leading the cart back to the rickyard was quite easy. We walked on the nearside of the horse next to the hedge, holding the ring at the end of the horse's bit or the rein close to it. The only problem was leading through the gateways. These were not much wider than the carts and usually rutted and uneven. Most of them opened onto narrow lanes or tracks. The horses tended to try to rush the gateways to use their momentum to carry them through and had to be held back. You had to ensure that nothing was coming and then lead the horse through, checking that it was dead in the centre of the gateway and that it continued straight on until the cart was clear of the gate posts. By this time, the horse's head would be almost in the hedge at the far side of the track and it would be fighting to begin the turn. Not resisting this and allowing the horse to start to turn too soon and cut the corner, usually resulted in the cart hitting a gate post and probably hitching it out. I can boast that I never took out a post, although I marked one or two where a hub grazed it.

The procedure was exactly the same when towing a loaded trailer with a tractor, and similarly the bonnet of the tractor had to be right at the hedge at the far side of the track before the turn was commenced. The axle on those early trailers was at the centre of the trailer. Thus, when the tractor was steered to the left, the front end of the trailer moved towards the right and the rear to the left, an idiosyncrasy that often had disastrous implications for gate posts. Undoubtedly, the modern trailer with the axle at the rear is easier to reverse and to drive through narrow spaces.

On arrival at the rickyard, the cart was pulled as tightly as possible alongside the base of the rick being built. Then the rope securing the load at the back was untied and thrown forward over the load. Then I would climb up onto the shaft, step up onto the horse's back, then up onto the front lade and from there scramble up to the top of the load using the rope to pull myself up. The rickmaker would hand me a pick so that I could begin to unload.

The crop from the larger fields was often enough to build an individual rick. Each rick was built on a layer of *vaggots of 'ood* – bundles of brushwood, each tied with a withy band, to help to keep the base of the rick dry.

The procedure for making the rick was similar to making the load on the cart but on a much larger scale, each row of sheaves overlapping the previous one to tie it in. The sheaves were pitched from the cart two at a time to the rick maker, or to his assistant if the rick maker was working at the far side of the rick. As the rick grew higher, the sides were allowed to slope slightly outwards to prevent rain running down the sides and into the butts of the outer rows of sheaves.

The rick maker relied on reports from the field on how much of the crop was left to bring home to determine when to start building the gable top of the stack. At this point the pitcher would be pitching most of the load uphill and ultimately, as the rick neared its peak, to the rick maker's assistant standing in a space left for him at eaves height. The helper would scoop the sheaves off the pitcher's fork with his own fork and hoist them to the rick maker at the top of the rick.

The outward sloping sides meant that ricks could be unstable, so one or two long, straight tree branches with a short fork left at the end were pushed high up into the side of the rick, their other ends jammed into the ground, to prevent the ricks leaning. These rick props were selected and cut when hedging during the winter. The rick gradually settled over the first few days, and so the props had to be reset from time to time.

Once the rick had settled, the sloping ridge top was thatched to keep the rick dry. A *haler,* or tarpaulin, stretched over the top and lashed down kept the rain out until the rick was thatched. Combed wheat straw which we called *reed* was used for thatching the ricks, as well as the farmhouse and the two largest barns. (Articles about thatching sometimes refer to it as Devon reed.) First the large bundles, or *niches*, of reed were immersed in a horse trough to soften the reed and make it supple. The reed was then spread out in layers, butts down, along the slope of the ridge working upwards from the eaves, each layer overlapping the previous one. The reed was held in place by thatching ropes stretched out along the reed and pegged down into the sheaves below with 'spars'. Thatching in this way would keep the rick dry until the winter when the corn was threshed.

The spars were made from hazel sticks cut from the hedgerow when hedge laying, or from the wood in the middle of the farm during the previous winter. The sticks were quartered using a spar hook, each length then being pointed at each end and twisted into a U-shape like a hairpin. One end of the spar was gripped in each hand and by simultaneously twisting and pulling the ends together the required shape was formed. The spar hook used for quartering and trimming had a short straight blade with a right-angled hook at the end. The blade was driven into the end of the hazel stick and twisted from side to side to work the split carefully down the full length. The point at each end was formed with three accurate cuts with the hook. We sometimes used the same tool for cutting out thicker branches when hedging in wintertime.

As on the ricks, reed was used on thatched houses in North Devon. However, the reed was laid on much more thickly, each layer almost completely overlapping the one below. The bottom ends were tapped into place to form the required slope with a 'leggat', or 'batter', so that only the very bottom ends of the stalks were visible. As a result, the straight hazel spars that were used instead of thatching ropes for durability were hidden. This construction resulted in a thick roof that was both warm in winter and cool in summer and would not need to be replaced for upwards of 40 years. Sometimes, the roof would be covered with chicken wire to prevent bird damage.

Carting could only be done when the weather was dry, and I have good memories of enjoying a picnic tea whilst sitting on a pile of sheaves on warm, often very hot, sunny days. The tea would be brought to the field in a wicker basket on one of the returning carts. There would be sandwiches of egg, tomato or cucumber, all produced on the farm, washed down with cold tea in bottles or in the milk cans with lids used on the morning milk round.

George had built himself a large greenhouse in the walled garden while he still worked for his father. Here he grew tomatoes and cucumbers very successfully, the surplus being sold on the milk round. Rationing continued for several years after the war had ended although manual workers were allowed extra rations of items such as cheese. I had these extra rations when I was working at my second farm, and my sandwiches for both my mid-morning break and lunch were regularly cheese with home-made pickle.

Teatime was a welcome pleasant break but as soon as we had finished eating the farmer was on his feet ready to start again. At the end of the day, however, if the last load was only a small one, we could ride home and relax, lying back on the sheaves, lulled by the motion of the cart, the rumble of the wheels and the clip-clop of the horse's hooves. It was often late in the evening before we finished, and the setting sun cast a red glow over the landscape. It was pleasantly cool after the heat of the afternoon. For a teenager, they were carefree idyllic days during the long summer holidays from school.

However, it wasn't always like that. There were wet summers when we all worked furiously during brief dry spells to cart the crop home against a background of threatening skies, heavily laden with moisture, often having to break off and return with half loads if it began to rain. Then there was a scramble up ladders to the top of the half-finished rick to struggle to unroll and lash down the heavy haler that someone had carried up the ladder on his shoulder. This was often opposed by the wind that often accompanied heavy rain and threatened to blow both the haler and us off the rick. By the time it was safely secured, we were wet through. There was never time to don coats but somehow it never seemed to deter us, even though next day it might all happen again; and often did!

Although there were many days when harvesting was interrupted by rain, the worst by far that I can recall was the day preceding the Lynmouth floods in 1952. I was home from university on vacation and, because it turned out to be such a dreadful day, I can remember it clearly. It was dry in the morning and although the sky was dark and foreboding after lunch we decided to cut corn. While we were preparing the binder in the barn at the corner of the road by the farm entrance, the heavens opened and the rain poured down. I have never experienced such a deluge in this country since. Although we were used to walking around and working in the rain, it was so heavy that we stayed in the barn for ages hoping that it would slacken, as it was clear that we should be drenched making the short dash to the farmhouse. It didn't, and for the only time that I can recall it was decided that, apart from milking, work would end for the day and everyone would go home. The rain continued unabated all evening, accompanied by heavy thunder and lighting, and we lost our electricity supply for much of the night. Next morning dawned fine, and it was only later in the day that we began to hear the news of the appalling disaster that had struck Lynmouth and Barbrook.

Chapter 3

Potato Harvesting

ONE or two fields of potatoes were grown each year on both farms on which I worked. They formed a useful 'break' crop within the crop rotation programme that was practised. Crop rotation was one of the few methods then available to reduce pest, disease and weed problems that could, and often did, reduce crop yields considerably. Continuous cropping with one type of corn or root allows associated pests and diseases to multiply, together with the weeds which flourish in competition with it. Nowadays, the farmer tends to rely on an extensive range of pesticides to deal with all such problems but in the 1940s chemical controls were in their infancy.

Crop rotation has been practised with considerable success for several hundred years. Without the host crop to live on during a rotation break, pests and diseases decrease; and the different cultivation programmes associated with each crop in the rotation reduce weed populations. Weeds in root crops like potatoes, grown in rows wide enough to be able to walk between them, can be controlled by hoeing them out by hand or by using a mechanical cultivator. Potatoes were considered to be a 'cleaning crop' because, in addition to being able to hoe between the rows in the early stages of growth, later on in the season the haulm grows so luxuriously that it covers the ground and smothers any further weeds that germinate.

The one spray commonly used was Bordeaux mixture which contained copper sulphate and was used as a control for potato blight. This is quite a common fungal disease, particularly in the West Country, as it favours a warm, damp climate. The haulm is infected first and dies down but the fungal spores wash down into the soil and can infect the tubers. The disease in the tubers causes soft brown patches and they soon rot in storage. Spraying against blight began in July or August and had to be repeated at regular intervals. This disease was the cause of the potato crop failures in Ireland in the 1840s which resulted in severe famine and caused so much misery and extreme poverty. It also exerted an influence on world history as it led to the massive emigration of Irish people, with vast numbers of them going to live in the United States. Recent research has indicated that the strain of fungus which causes blight in our crops today is different to the one that caused the devastation in Ireland.

Potatoes are grown in ridges. We 'mucked' the field with farmyard manure from the buildings where cattle were overwintered. It was then ploughed in and the soil cultivated to a fine tilth in which the ridges could be drawn. These were made by a ridging plough pulled by two horses. This resembled an ordinary horse plough in shape and construction, but a ridging body was attached below the frame of the implement instead of a ploughshare and mouldboard. The ridging body resembled two curved mouldboards set at an angle to each other and joined at the front. When this was pulled through the soil it produced a vee-shaped furrow, pushing the soil aside to form a half ridge at each side of the furrow. When the next furrow was drawn parallel to the first, the plough completed one ridge and produced the next half ridge. The method has not changed, although now a tractor will pull several ridging bodies and plant the seed potatoes automatically at the same time.

We planted seed potatoes by hand. The seed tubers were carried in a wooden chip basket set down on the ridge. The planter took a handful of seed tubers in one hand and walked along the adjacent furrow placing a tuber in front of each foot, matching the length of each step to the seed spacing required. Then the basket was lifted with the other hand, moved forward and put down again, and the planting continued. This was back-aching work, especially when the basket was full. It became easier as the basket was emptied, but then it had to be refilled and the process started all over again.

As soon as several rows had been planted, the ridges were split by the same ridging plough used to make them, the soil flowing over the planted tubers and forming new ridges over them.

Soon after the first leaves had appeared the ridging plough was pulled through the valleys to build up the ridges and remove any weeds that had germinated. This operation might be repeated again at a later date to ensure that there was plenty of soil over the growing tubers. It was important to ensure that the ridging up was done thoroughly, as any tubers that grew out of the side of the ridges turned green, and green potatoes are harmful to eat. Once the crop had covered the ground there were no further problems with weeds.

The farm's workers were allowed to have a row of potatoes for their own use. There was no demarcation when cultivating and fertilising the crop, but the worker was expected to hoe any weed from their own rows and to plant and harvest them, usually with the help of wife and children.

If a small quantity of potatoes was required at harvest time, either to sell on the milk round or for use in the house, they were dug by hand with a potato digger. This had a long handle with two short and stout flat tines at right angles to it. It was held with both hands and driven deep into the ridge at the far side of a root and then pulled back towards the worker. With practice a whole root of potatoes could often be pulled out at one go ready to be picked up, although the whole area would be dug over with the digger after the potatoes had been collected to ensure that no potatoes were missed.

The bulk of the farm crop was harvested mechanically by using the ridging plough again, this time to split the ridges and expose the tubers. We split one ridge at a time and picked up the potatoes before continuing. These were collected into the same chip baskets used when planting and then emptied into sacks. Splitting the ridges was not an ideal method because many potatoes were barely visible or still buried and had to be grubbed out by hand by the pickers. This slowed down the process, and undoubtedly a proportion of the tubers was missed. Apart from the loss of crop, a few of these would grow next year and become troublesome weeds.

Potato spinners and, following after them, elevator diggers, were more effective in lifting the crop and leaving it fully exposed on the surface. When these machines were used continuously, large numbers of pickers were required. In parts of the country where large acreages of potatoes were grown, gangs of itinerant workers who would return year after year to the same farm were employed. Present-day machines are usually complete harvesters. They scoop the potatoes out of the ground and convey them to a hopper on the harvester in one continuous operation. There is sometimes provision for clods and stones carried up with the tuber to be removed by workers riding on the machine. Alternatively, de-stoners can be used to remove stones from the ridges before planting and, if care is taken when cultivating to prevent the formation of clods, there is no need for workers on the harvester. Any clods or stones that do pass over the machine with the potatoes can be removed when the crop is riddled before sale. The complete harvesters cost many thousands of pounds but cover the ground very rapidly and require very little labour to operate.

On my second farm, where I worked full time after leaving school, we used a spinner. This was designed to be pulled by two horses but had been converted to enable it to be drawn by a tractor. The potatoes were spun out of the ridge by pairs of curved steel tines passing through the ridge from one side to the other. The tines were mounted around the rim of a rotating steel wheel that was driven by the two land wheels that supported the machine.

It was while potato harvesting at this farm that I had my first and only experience of working with a gang of land girls. The regular farm work force was just my employer and I, supplemented at busy times by part-timers, and so my boss had arranged for a gang of Welsh land girls billeted nearby to help us lift our crop. He and I operated the tractor and spinner and would spin out several rows that the girls then picked up. As the spinner had been designed as a horse-drawn machine, the control levers could not be reached from the tractor and could only be operated from the seat that projected from the rear of the machine. My boss drove the tractor and I rode on the spinner. My job was to adjust the working depth of the spinner to ensure that the potatoes were all spun out without the tines going too deep and burying them again.

The girls worked in pairs and each pair was allocated a length of row to pick up. They put the potatoes into chip baskets which when full were tipped into sacks. One girl would hold the sack open while the second carefully emptied the basket into it to avoid bruising the tubers. Although the spinner exposed most of the *teddies*, as we called them, a few would be barely visible and had to be grubbed out by hand. Picking up the potatoes was another back-aching job, not unlike planting them, as the picker was bent double all the time and the basket had to be moved forward regularly; and it soon got heavy. In between operating the spinner my boss and I would load the bags onto the trailer and cart them back to the farm. When we had any spare time we picked up as well, helping out any of the girls who were falling behind.

I found working with these girls quite difficult. I was just over 18 but I was shy and had had little experience of the opposite sex. I had attended a boys' grammar school that was a totally male environment. It was still wartime when I started there, but nonetheless all the teachers were male, most of them beyond the age of military service, although for a brief period a French woman was on the staff to take classes in her language. Although the girls' school was on an adjacent site, a high fence separated the two schools and boys and girls were not allowed to mix. When we travelled by bus to other senior schools in the area for sixth-form conferences, a double-decker was booked and the girls sent upstairs while we travelled on the bottom deck. Very few of my contemporaries at school admitted to having a girlfriend, even when we reached the sixth form. However, a number of end-of-term parties were organised by pupils from the two sixth forms. They were held in a private room in the town, and the evenings were largely given over to party games interspersed with one or two simple dances, all of which the present generation of young people would consider to be quite childish. Nor do I recall alcoholic drinks being served.

I lived more than two miles from the school at the very edge of the town, surrounded by fields. No girls of my age lived near me and, for that matter, only one boy. There were three or four girls close to my age at the chapel we attended who were also members of the youth club to which I belonged, but we all went out as a group and they each had attachments. A land girl did work for a time at my home farm while I was in my mid-teens, mainly helping with the milking. Lilian was much older than me and I thought of her just as a mate, working side by side with the rest of us as one of the team and getting just as dirty and dishevelled.

As a result, I found working with a gang of about a dozen young women not much older than I was, disconcerting and quite unsettling. Whenever I was near them, they always seemed to be giggling and whispering comments loud enough for me to overhear. Usually, they seemed to concern me and particularly my appearance and physique. I guess that in today's era of political correctness it would be regarded as sexual harassment.

Occasionally, my employer had to attend to other work and I was left on my own with the women, nominally 'in charge'. To my innocent surprise, whenever I needed assistance to ride the spinner while I drove the tractor, or to load sacks, it was always the same dark-haired girl who was right next to me and eager to help.

The experience of running the operation when the boss was not there should have increased my confidence, as I managed to keep it running smoothly and he always seemed satisfied with progress when he returned. However, an incident midway through the week shattered any new-found confidence and caused me great embarrassment.

The seat on the spinner was typical of those fitted to many horse-drawn machines and early tractors. It was a shaped bucket seat made of cast iron, with holes in the bottom to allow rainwater to drain through when the machine was left in the open. All such seats were hard and unyielding, and the perforations made them particularly uncomfortable, so a sack was usually put on them to make them bearable. Cast iron is brittle, and several of the drainage holes in the seat on our spinner had been damaged, and the bits that had broken away had left jagged edges. The sack also covered these. On that memorable day I had forgotten the sack to put on the seat and, inevitably, when I dismounted I ripped the seat out of my trousers. Of course, the girls immediately spotted this and the howls of laughter amounting to near hysteria that resulted at this trivial mishap followed me and left me with a very red face as I slipped away back to the farm. There the farmer's wife just managed to keep a straight face and repaired them for me. She was younger than her husband and not much older than the land girls, but she was always very kind to me and often slipped out to the shippon, or cowhouse, with a cup of tea in the afternoon while I was milking.

I was not sorry when the field was finished and the girls departed, including the dark-haired one who had seemed to lose interest after the seat incident.

Chapter 4

Horse Power

IN the late 1940s, heavy draught horses were still in common use on farms all over the country, although tractor numbers had been rising during the pre-war years, particularly in the arable areas of eastern England and Scotland. Adoption of the tractor in North Devon had been much slower. The tractors of the time were not suited to the small, steep fields and the wet, heavy soils characteristic of the area. As a result, horses were still widely used for the majority of farm operations.

The horses we used were not pure bred. They tended to be slightly lighter in build than the pure-bred Shire, the largest breed of English draught horse, although undoubtedly there was a great deal of Shire in them. It was argued that our horses were strong enough for the work in our small fields and narrow lanes that limited the size of carts and implements and they had the advantage that they were quicker on the move than the more ponderous Shires.

Improvements in tractor design and difficulty in obtaining labour, particularly at the more remote farms in the area, contributed to the demise of heavy draught horses in North Devon. Modern four-wheel drive tractors with differential lock, good turning circles and sophisticated hydraulics which provide weight transfer and assist wheel grip are well suited to the area and, sadly, the horse no longer appears to have a niche.

At my home farm there were four horses and a Standard Fordson tractor when I started to spend time there, although there had been a greater number of horses before the tractor arrived. Their names were Charlie, Darkie, Flower and Punch. Punch was the smallest and lightest of the four, and his principal role in the morning was to pull the trap that was used as a milk float, although at harvest time he also did light carting in the afternoon. Flower was a chestnut with black mane and tail. Punch and Charlie were both brown with dark brown manes and tails and, of course, Darkie was black. All had short tails as it was accepted practice at the time to dock the tail and to keep it short.

At first, Flower was the only mare, the rest being geldings, but when Punch was retired a mare called Trixie replaced him. A van was acquired to do the milk round shortly after and so she was not with us for long. The farm staff thought that mares were more temperamental than geldings, particularly when in season, and as a result Flower was not as popular as the other horses. Maybe that is why, as the youngest worker and newest recruit, I found myself regularly partnered with Flower when I became old enough to take a cart when we were harvesting. If she was a typical example of a working mare, their distrust of them was well founded. Nevertheless, I became very fond of her, probably because we worked together so frequently that I thought of her as my partner. However, she could be a handful and over the years she literally left her mark on me and got me into difficulties on a number of occasions.

When I started at the age of 11, Flower, like the other field horses, seemed huge. She was probably about 15 hands high, that is 60 inches (just over 1.5 m) measured up to her withers at the base of the neck, so she towered above me and I had to reach up to her bridle to lead her. This led to my first mishap with her, and I still bear the mark.

George and I were hoeing a root crop. He was doing the skilled job of guiding and controlling the hoe and I was leading Flower. It was not usually necessary to lead the horse once the crop was a couple of inches high because a trained horse would walk along between the rows with little or no guidance and not veer onto the rows or tread on the plants. However, ideally the first hoeing was done just after the crop had germinated to remove any seedling weeds that had germinated between the rows in order to reduce competition and to give the plants the best possible start. At this stage, it was difficult to see the rows clearly and so it was common practice to lead the horse.

At the end of the drill the horse had to walk right up to the hedge to ensure that the hoe worked right to the end of the row. A narrow 'headland' where the drills stopped short of the hedge was left at each end of the field to enable the horse and hoe to be turned around. It was my job to ensure that Flower went right to the hedge and then to guide her around in a U-turn ready to start back down the next row. This had to be done slowly to give George time to lift the hoe out of work, pull it around by hand and reposition it in line with the next row. When Flower reached the hedge and stopped

I had to walk backwards, pulling her bridle to bring her half round and towards me, still keeping her nose close to the hedge. Then, I had to move across in front of her, or dodge under her head and neck, to grasp the other end of her bit and finally complete the U-turn on her other side. Flower knew the procedure as well as I did but that day apparently decided to show off by doing it all on her own without stopping and waiting for me. I was totally unprepared for this and she was on top of me before I realised what was happening, and I couldn't back away fast enough to avoid being trodden on. Fortunately, the cultivated ground was soft or it could have been much worse, although I still have an enlarged toe joint to remind me of her.

Flower's biggest problem was her intense dislike of flies, especially horse flies. These blood-suckers with elongated bodies were common in warm summers and were a particular affliction during harvest. I haven't seen any for years. Possibly the reduction in the horse population has influenced their numbers, although the big increase in pesticide use may also have had an effect. Despite the name they were not fussy about species and were quite happy to adopt me and my co-workers as alternative hosts. They pitched on any exposed skin, and often a sharp prick was the first indication that one had landed on you although they were quite large. Even if they were squashed immediately or quickly slapped away they had always drawn blood.

Perhaps Flower was thinner-skinned than the other horses as they seemed to be less bothered, even though the flies pitched on them just as frequently as they did on her. As a result, she would never stand completely still in hot weather and restlessly moved from one foot to the other. She would toss her head up and down and make her skin ripple to prevent flies landing on her or to dislodge them. As a result, she was very difficult to lead. Avoiding her big feet was my main concern as she constantly lifted her front legs as high as she could to try to dislodge flies from her chest with her knees; and being trodden on was not an experience I wanted to repeat.

Her inability to standstill did have a more serious outcome when I was older and had progressed to making the load on the cart. It was always difficult to maintain one's balance when Flower was between the shafts because the cart was always on the move as she edged about. On this occasion the load was almost finished. I was standing right on the top and the pitcher was at full stretch, so I had to lean forward to pull the sheaves off the tines of his fork. As I did so, Flower moved forward. I momentarily lost my balance and fell forward onto one of the fork tines that was projecting through the sheaves, the tine penetrating deep into the soft flesh of my palm. The scar is still visible today as another reminder of my love-hate relationship with her.

When returning her to the stable, Flower would always stop just before entering and raise her head to its full height so that she couldn't pass through the doorway. She just stood there, as if to say, 'It's dark in there and can't you see that I am too tall to go in through the door?' After what seemed like ages of cajoling, pleading and threatening she would suddenly decide that she had made her point and would charge through the door, hooves clattering and sliding on the brick floor, straight into her stall which was opposite but to one side of the doorway. I soon realised that the safest plan was to let go of her bridle or halter rope and keep out of her way until she was in.

Darkie was slightly smaller than Flower but more powerful and determined. This was brought home to me when George and I took Flower to a smallholding about a mile from the farm to haul the carcass of a dead animal out from the stable there and up a short but steep drive to the gate. There, it could be collected by the fellmonger with his wagon. We attached the draught chains we had taken with us, and Flower made one unsuccessful attempt to move the deadweight but then refused to try again. She had made up her mind that it was beyond her, and that was that. No amount of persuasion would shift her. We took her back to the farm, fetched Darkie and hitched him up instead of Flower. He leaned forward, dug in his hooves and walked the carcass straight out of the building and up the slope to the road without stopping.

I must add that I never saw or heard of any horse being ill treated on the farms near us, although perhaps it did occur on other farms. They were urged on by talking, by pulling on the bridle or by flipping the reins across the back if they were being driven. Maybe I was lucky but the seven or eight horses that I worked were all willing and reasonably behaved. In my experience, they only jibbed or refused if they knew that the work was beyond them and, if so, there was not much point in trying to force them and possibly induce a strain or worse.

On TV and on film horsemen frequently say, 'walk

on' to start their horse. I never heard any farm worker say that. I copied my fellow workers and I can only remember calling, 'cum on', using a clicking noise with my tongue and flipping the reins to start the ones that I drove and saying, 'whoa' and pulling back on the reins to stop. The skilled horsemen also had commands to turn left or right. One of these was, 'cum hither'. I think that this was to turn left but I am no longer sure.

Although Darkie was a strong horse, he had his dark side. He had an intense dislike of being made to go backwards. This always surfaced when he was being backed in between the shafts of a cart. I always believed that he had been backed into an obstruction that had hurt his leg at some time and was wary of it happening again.

Normally, the procedure for backing a horse between the shafts was to hold the ring at one end of the bit or the rein adjacent with the near hand, whilst facing backwards towards the cart. This enabled me to guide the horse backwards, taking care to avoid getting either of the back legs caught against a shaft. When this was done with Darkie his head would come around and he would grab at your forearm with his teeth. I had been warned of this and told to watch for his ears to go back, as this was a sure sign that he was about to bite. The safe method to back him was to stand in front of him and hold both ends of the bit so that he could not turn his head far enough to reach your arm, although this made backing more difficult. It was very easy to forget this unfortunate habit, particularly when he was at work pulling a cart and it was necessary to back it.

Charlie was a much older horse and everyone's favourite. He was close to retirement and was not worked as frequently as the younger ones. Although he was considered to be a very steady horse, he had an intense dislike of crossing the bridge over the railway close to the station. There was a siding by this bridge where a tank engine often stood. Trains from London via Exeter were often divided at Barnstaple, and the engine waiting to take on the Bideford portion used to stand in the siding. The engine would usually be letting off steam that rose in clouds and sometimes drifted over the bridge. It was this that seemed to terrify Charlie. He had to be held very firmly with both hands on the bridle while he tried to back away, and he usually crossed the bridge sideways so that he could keep the steam in view at all times, much to the consternation of passers-by. Fortunately, the road was not as busy then and drivers were more used to meeting horse traffic.

We rarely had to go past the station but there was no alternative when we had to take Charlie to the farrier to be shoed. I can only recall taking him once on my own. I rode him as far as the bridge and dismounted to lead him across as I had been instructed, but after crossing the bridge I couldn't climb back onto his back again because he was too tall. We had no riding saddles with stirrups for the carthorses and just put a sack on their backs to sit on, so it was almost impossible to climb back on without some form of mounting block. I walked him the rest of the way to the farrier's shop, which was not too far, but it was much further to walk after crossing the bridge on the way back. However, at the foot of Sticklepath Hill, not far from the bridge there was a milestone and I managed to edge him close to it so that I could climb onto it and scramble onto his back from there.

At my second farm, where I worked for 15 months full time, there were two horses and two tractors, a Fordson Major and a Ferguson TE20. Unfortunately, both horses, one chestnut, one black, were called Prince! This could be awkward when working them as a pair if it was necessary to encourage one that wasn't pulling as hard as his partner or to give an instruction to just one of them.

As there were two quite new tractors on the farm, the two horses were not worked as frequently, although they still did quite a bit of field work, particularly in soft soil when my employer was keen to avoid using the tractor which would leave ruts. Both tractors had rubber-tyred wheels and so much of the haulage was done with them. However, it was often more convenient to move small loads with a horse and cart, and it was while carting such a load that I had my worst experience with a horse.

My favourite of the two Princes was the black one, and he was between the cart shafts one day as we left the farmyard and went up the steep drive that led to the road in front of the farm. A horse climbing a slope and pulling a load angles the fronts of his hooves downwards so that he can dig in the front edge of his steel shoes to get a grip. The drive was tarmacadamed and the surface was smooth, and that day it must have been slippery as well. When we were part way up, his front feet failed to get a grip and slipped back under him so that he went down onto his knees, and the weight of the cart began to drag him back down the slope. I managed to find a stone to wedge behind the wheel of the cart, and with some difficulty Prince managed to regain his feet. Fortunately, the load was evenly distributed over the central axle of the cart and so there wasn't much weight on the shafts. If there had been, he would have been unable to rise and somehow I would have had to unharness him where he was. Even so, his eyes were rolling and he was very agitated. I don't know whether my eyes were rolling but I was certainly scared. Luckily, his knees were not grazed or damaged, and I have to confess that I never told my employer what had happened. I never took a laden cart that way again. Instead, I used an alternative track from the rear of the farm buildings that was not surfaced with tarmac.

Sadly, black Prince developed a condition called ring bone in his foot that the vet considered incurable, and some time after I left the farm I heard that he had been put to sleep to prevent him suffering.

For most of the year when the horses were not working, they were released into a field where they could graze. It was always a field near the farm so that it wasn't far to take them after a day's work or to collect them next morning. They loved to be let out, particularly the first time in the spring after being in the stables all winter except when working, and they would gallop around, throw up their heels and roll on the ground. A big horse rolling with all four feet in the air was quite a sight. Surprisingly, even after a long day's work they always seemed to be able to summon up enough energy for a short canter before settling down to the serious business of eating.

They were not always as keen to be caught and brought in again, especially if they had been out all weekend. Their usual game was to wait until you had walked up to them and then to turn away for three or four steps and wait for you to approach them again. This game could be played several times before one or other would give up. Once one had been caught the others would usually follow, even without being haltered. Attempting to run after them was a waste of time because they would enter into the spirit of the game and dash off to the far corner of the field at a gallop.

The horses did not wear halters in the field and so, when you had caught one, you had to reach up and grasp the forelock between the ears, and then slip the woven webbing halter around the muzzle and up over the ears. Catching the two horses at my second farm was particularly tedious because the field opposite the farm gate where they always grazed was so steep that it was a struggle to walk up it, let alone chase after a couple of horses.

Flower had learnt how to open a field gate and could let herself and her companions out into the next field, or worse, out onto the road. Fortunately, there was little traffic on country roads then and so this was more of an irritation than a danger as she usually wandered home and down to the stable. The catches on the gates weren't easy to open. They consisted of an L-shaped latch, the longer arm being loosely linked to an eye fixed to the gate. The shorter arm was dropped into a U-shaped staple fixed to the gate post, and ended either with a loop in which a padlock could be inserted to secure the gate, or with a barb like that on a fishhook. Both prevented the latch from being undone by simply lifting the latch. Nevertheless, Flower had mastered the skill of unlatching the gate and pushing it open.

At my second farm we made the mistake of leaving it too late in the year before keeping the two horses in the stable overnight. Unexpectedly, one night early in winter it snowed, and when I arrived late for work following a long and hazardous cycle ride along slippery roads, I was to discover that the horses were not in their field but somehow had broken out.

After searching the nearby fields and the roads roundabout for some time without success, my employer received a phone call from a neighbour informing him that both horses were at his farm a couple of miles away. My boss took me in the car to collect them and left me to ride one and lead the other back to the farm. I had taken a bridle and reins for the first horse and a sack to put on its back to sit on, and a halter for the second one that was to be led.

We set off along the slippery road and both horses started to slip and slide, particularly where the road was steep. They seemed to enjoy this, so much so that I realised that the one I was riding was deliberately leaning back to induce a slide when going downhill. Whenever this happened I was pulled backwards by the tension on the halter rope from the horse behind and almost unseated, only to be jolted forwards again when the second horse cannoned into us as he slid after us. After a while I decided to get off before I fell off, and to lead both of them all the way back to the farm.

Riding a carthorse without a saddle or stirrups and

with only a sack to sit on is easier than one might think. Their backs are very broad, and so straddling them causes one's feet to stick out at each side, making conventional riding techniques difficult. However, being so wide they provide a fairly comfortable and secure seat, although it is possible to feel their powerful back muscles undulating under you. But, as George's father used to say, probably not originally, 'Second-class riding is better than first-class walking.'

For much of the year, apart from winter, the horses spent most of their time out in the field when they were not working. There was usually enough grass to make it unnecessary to supplement their diet unless they were doing particularly heavy work. In the winter they lived in the stable. Although it was convenient to have them on hand, it entailed extra work because they had to be fed and cleaned out twice a day and taken out to the water trough to drink because there was no water supply in the stable. Of course, all this had to be done seven days a week.

They were fed hay stored in the stable 'tallet', the loft above the stable. There was a hayrack at the front of each stall over the manger, and hay was forked down through holes in the floor of the tallet straight into the hayrack in front of each horse. At harvest the hay had been pitched up into the tallet though a door at loft floor level. The tallet was reached by climbing a vertical ladder fixed to the wall that led to a hole in the floor of the loft.

Sliced mangolds were put into the manger in each horse's stall, and rolled oats, stored in a wooden chest at the back of the stable behind the stalls downstairs, would be fed to the horses when they were to undertake very heavy work.

Each horse in a stall wore a halter, and the halter rope was threaded through a ring at the front of the manger and then through a hole in a wooden ball and knotted. This allowed the horse plenty of free movement but the weight of the ball automatically took up any slack in which the horse might have become entangled.

Letting them drink involved taking out each horse in turn across the yard to the water trough. Of course, taking Flower led to the usual pantomime when she had to be taken back into the stable. However, being a mare she was easier to clean out as all her wastes were deposited in the shallow channel that ran behind the stalls. The stalls were narrow, so you had to lean against the rear quarter to persuade a horse to move to one side so that you could slide along beside it to put food into the manger or replace the bedding straw. Often the horse would move over again while you were at the manger, and you would have to push out past it, quite a frightening experience when one was young and still relatively small and they appeared to be so large.

The harness for each horse was hung on a wooden peg on the back wall of the stable. What they wore depended on what job they were to do, although they always wore a bridle with a simple straight steel bit in two halves loosely linked at the centre. If they were to pull an implement like a hoe or a ridging plough with trace chains (sometimes described as draught or pulling chains), the only additions were a horse collar and 'haimes', or 'hames'.

Thick and wide, the collar was made of leather but padded so that it would fit snugly and comfortably against the front shoulders of the horse without chafing and causing sores. The collar opened at the top and had to be pulled apart to allow it to be lifted up around the horse's neck. Once in position the two sides were allowed to spring together again and were secured with straps over a leather flap covering the join. In other parts of the country the collar was commonly in one piece. Being narrower at the top than at the bottom, it was necessary to turn this type of collar the wrong way up before slipping it over the horse's head and then to invert it so that it rested comfortably against the shoulders.

The hames consisted of a pair of steel bars curved to fit into matching grooves in each side of the collar. Held together at the top by adjustable straps, the hames were normally left alone once adjusted for a particular horse, and were done up at the bottom with a short length of chain and a simple clasp. The tops of the bars projected vertically well above the top of the collar. Apart from providing a convenient peg on which to hang one's overcoat or lunch bag, I never did find a reason for this extension. A flattened hook hinged to each hame provided the attachment points for draught chains.

An additional harness was required if the horse was to pull a cart or implement fitted with shafts. This consisted of a saddle, secured with a belly-band (a strap that buckled under the horse), and breeching that consisted of a wide horizontal leather strap passing around the rear quarters and suspended at each side from a strap running from the saddle along the back towards the tail. This all had to be put on as one unit, the heavy saddle lifted with one hand and the breeching and associated straps with the other. The saddle was made of leather and padded like the collar. Two ridges curved sideways across the saddle and a double chain with interwoven flattened links fitted in the groove between them.

To attach the cart the horse was backed between the shafts. These were lifted so that the back chain across the saddle could be connected to hooks on the shafts to support them. If you were on your own, it was done by lifting up the shafts and attaching the chain at one side before diving under the horse's neck, whilst still supporting the shafts, to hook the other end of the chain to the other shaft. Next, short chains from the hames were attached to forward-facing hooks mounted on the shafts to enable the horse to pull the cart. Finally, two further short chains attached to the horizontal breeching strap were connected loosely to rear-facing hooks on the shafts. These enabled the horse to move the cart backwards by pushing against the breeching with its hindquarters. Although not part of the horse's harness, a loose rope between the shafts under the horse's belly limited the amount that the shafts could rise up if a heavy item was placed right at the rear of the cart.

We always used a rope for reins. This was tied at each end to the rings at the ends of the bit with a simple knot that I called a 'rein knot' and which I have never seen used in any other context. I still find it very useful if I want to form a loop at the end of a cord or rope.

Although there were horse brasses hanging on the wall in the stable, none of our horses were ever fitted with the breast straps on which they would be attached and so I never saw them in use. Horses at shows with all the harnesses clean and the brasses polished and shining make a fine sight but our horses were essentially work animals and there was no time for frills.

Chapter 5

Tractor Power

THE Standard Fordson tractor that was pulling the binder when I wandered into the cornfield for the first time, was a late example of the model N that first appeared in the late 1930s. Our version was characterised by narrow, almost non-existent wings curving outwards over the wheels. Earlier examples had wide, flat wings. In most other respects it looked not unlike the first Fordson made in 1917 at Dearborn in the USA, although many mechanical details had changed in the meantime. Primarily because of competition in the States, Fords moved all their production to Cork in Ireland in 1929, and to Dagenham in Essex in 1933 where the factory built there became a major site for the mass production of Fordson tractors. The tractors were called Fordsons, rather than Fords because another small tractor manufacturer was already using the name Ford in the US when Henry Ford wanted to start production.

Despite the improvements over the model F, the original tractor of 1917, the N was still a crude and not particularly efficient machine. It had no electrics and could only be started by cranking the engine with a starting handle, although in this respect most other tractors were the same.

The engine ran on tractor vaporising oil, TVO for short, a less refined form of paraffin. This was not taxed and as a result was much cheaper than petrol. Less volatile than petrol, the engine had to be started on petrol and switched over to TVO once it was warm. A fuel tank mounted along the tractor above the engine was divided into two with a small compartment at the rear for the petrol. There was a tap under the tank to switch from one fuel to the other. A blind in front of the radiator could be rolled up to help the engine to reach working temperature more quickly. A water temperature gauge on the top of the radiator gave some indication of when it was safe to switch over. If this were done before the engine was properly hot, the plugs would 'oil up' with unvaporised fuel and would have to be removed and cleaned before the engine could be restarted. Despite all this it was a reasonable starter, although quick starting first thing on a cold winter's morning could be ensured by taking the plugs out and putting them in the oven for a short while.

There was a very precise starting procedure. First, with the fuel tap turned off, the TVO in the float chamber had to be drained out through a drain cock at the bottom. Then the fuel tap was turned to the 'petrol' position so that the engine could be started on it. To avoid this initial stage we switched over to petrol and ran on it for a period before turning off the engine – if we remembered – so that the float chamber was already full of petrol when we wished to start it again from cold. Next the ignition was retarded by turning a small lever. Then, the procedure was to pull up the starting handle three times and, with any luck, it would start on the fourth half turn. It was important to make sure that when cranking the engine you kept your fingers and thumb on the same side of the handle to prevent the thumb being broken in case the engine backfired if the engine had not been retarded sufficiently. Finally, the ignition was advanced again and, once the engine had reached working temperature, the fuel tap could be switched to TVO.

The gear system on the Standard Fordson was unconventional and included a multi-plate clutch immersed in oil and a transmission brake, the final drive being provided by a worm wheel and gear. When the clutch was fully depressed the drive was disengaged and the brake applied as well, so the tractor stopped immediately. Even without pushing the clutch pedal down into the 'braking' range, the tractor still stopped almost as quickly because of the drag from the oil-immersed clutch and the worm final drive. The worst feature of this unusual transmission was that as soon as the gear lever was moved into neutral the friction drag from the oil in which the clutch plates were immersed caused them to begin to rotate again, even with the pedal still depressed. This made it very difficult to engage another gear without raking them noisily, particularly when the oil was cold first thing in the morning. It could be done by gently moving the gear lever until the teeth could be heard snickering against each other, and to use this resistance to slow down the gears until the gear lever could be pushed right in. The alternative was to depress the clutch, grasp the gear lever in both hands and slam it from one position to the next before the oil drag could take effect. Failure to do this quickly enough resulted in even worse gear noise, and one had visions of broken teeth flying around in the gear housing.

Fortunately, gear changing was not a frequent event as almost all field work was done in second gear. If it was necessary to stop to make an adjustment to an implement, the clutch pedal had a convenient hook that could be used to hold it down, still in gear, while the adjustment was made. Sometimes, to avoid the problem of engaging a gear, we started the tractor with the gear to be used already selected and the clutch pedal held down with the hook. Not perhaps the safest of procedures or an arrangement that would satisfy a safety officer but, of course, there were none then. First gear was normally used only for attaching implements and for a few exceptionally heavy draught operations, and third gear for travelling on the road. The only extra on the tractor was a pulley mounted on the side that we used to drive a circular saw and a grist mill.

There were two tractors at my second farm: a Fordson Major, model E27N that had replaced the Standard Fordson in 1947; and a Ferguson TE20, the popular 'grey Fergie' that is still to be found in use on smallholdings more than 50 years on. At first sight the Major bore some resemblance to the old Standard, having the same oval-section double fuel tank mounted over an engine that looked like its predecessor, but the Major was longer and set up on bigger wheels so that it was much taller than the tractor it replaced. However, there were major mechanical differences. The engine was larger and more powerful. The three-speed and reverse transmission had a conventional single-plate friction clutch, and crown wheel and pinion final drive just like a car so that easier gear changing was possible. Independent brakes on the two rear wheels made very tight braked turns possible, and if one wheel started to spin on a wet surface it could be braked, so that all the drive transferred to the other wheel. Even when both wheels were inclined to slip, rapid use of each brake in turn would often provide sufficient grip on each wheel alternately to get through a sticky patch.

Perhaps the Major's biggest advance over the Standard was that it was equipped with a hydraulic lift system that enabled implements attached to three links at the rear of the tractor to be raised and lowered by moving a lever. Attachment was done by pushing a hole in the end of each lower link over a matching pin projecting from each side of the implement. It was completed by threading a loose pin through the hole in the end of the top link and matching holes in a jaw fixed high up on the implement. All the pins were secured with circlips, and all the attachment points were in ball joints that allowed the linkage to move freely. Lateral movement was constrained by adjustable check chains. In practice, attachment was not always easy because the tractor had to be reversed and stopped with great precision to allow the links to be fitted to heavy implements, and this was quite difficult to do on rough uneven surfaces. Fifty years on this still remains the way that mounted equipment is attached to a tractor although connection has been made easier.

The Major also had electrics, not that they were of much use as the battery was fitted above the steering column where it was fully exposed to the weather. So the battery was usually flat and we still had to crank the engine to start it.

The Ferguson had the same type of engine, started on petrol but running on TVO once it was warm, but it had a four-speed gearbox and effective electrics. The engine was started by pushing the gear lever against a starter switch. Although much smaller and lighter than either of the Fordson models, it had a superior hydraulic lift system. This automatically transferred weight from an implement that was working in the soil to the rear wheels, giving it much better wheel grip and often allowing it to outperform the bigger tractors.

The simple but very clever control system for the hydraulics that made this possible had been designed by Harry Ferguson. It was originally incorporated in a Ferguson Brown tractor, model A, in 1936 or just before, and later in a Ford Ferguson, model 9N, in 1937. By all accounts David Brown, Henry Ford and Ferguson were all strong characters and so after disagreements with the others Ferguson began to build his own tractors in what had been the Standard car factory at Banner Lane in Coventry, in 1947. It is a measure of his genius that the tractors of today still incorporate the basic principle of the weight transfer system that he developed before the war. Other manufacturers unsuccessfully tried to design an equally good alternative or to copy it. This led to a lengthy legal battle with Ford who manufactured the Ford 8N that was almost identical with the Ford Ferguson 9N that preceded it. The argument resulted in one of the biggest payments of corporate damages up to that time.

Each university vacation I returned to the home farm

to work, and by the time that I finished there a Fordson New Major, E1A, had replaced the old Standard. These first appeared at the Smithfield Show at the end of 1951 and were the last tractors built at Dagenham. They were vastly superior to the Standard Major that they succeeded. They had efficient electrics and a high-low secondary gearbox that provided six forward and two reverse gears which gave much greater flexibility for both field and road work. The hydraulics system was an advance on that of its predecessor, and the other major progression was that the engine was a more powerful unit and a diesel. This did away with all the previous problems with dual fuels and its performance characteristics were much more suited to farm work, although for a time a dual-fuel unit was still available as an alternative to the diesel.

The New Major was a pleasure to drive in comparison with its predecessors, except that it did have one rather frightening idiosyncrasy. The engine would slog quite happily like most diesels, but if the engine was allowed to overload to the point when it stalled, occasionally it would restart immediately on its own but *rotating in the wrong direction*, throwing out exhaust and oil from the air cleaner inlet and revving uncontrollably. The only way to stop it was to select top gear and let out the clutch sharply to stall it. I never discovered whether this was a universal fault of the mark or confined to our particular machine. It was certainly frightening when it did occur.

High-speed diesel engine technology appears to have developed more quickly in Germany than in this country and the US. Not only did they fit diesels in their wartime aircraft but German tractors were also fitted with them long before they appeared in the UK. Until the introduction of the New Major, diesel engines were rarities in both British and imported American tractors. However, most manufacturers soon followed suit and switched their production to diesels. Ferguson offered a diesel option at about the same time as Fordson.

The one exception in Britain was the Field Marshall, manufactured in Gainsborough. It appeared as early as 1928 in prototype form as the Marshall with an unusual slow revving, single-cylinder diesel engine, the cylinder being horizontal, altogether similar to the German Lanz. The Field Marshall that went into production in 1945 fitted with the same type of engine was much more successful than its predecessor, particularly when equipped with a six-speed gear system as in the Series 3 that followed in 1949. It was quite powerful for the period but I recall that it was initially more popular in North Devon for driving threshing machines than for field work. Its single-cylinder engine meant that it bounced up and down when it was idling, despite a very heavy flywheel, and as a result had a reputation for wearing out drawbar pins. The Field Marshall's engine design made it very difficult to crank-start and it was unsuited to electric starting. It was therefore equipped with a unique starting procedure involving the use of an explosive cartridge very similar to a shotgun cartridge in size and appearance. The cartridge was inserted into its chamber, the covering cap screwed down over it and a protruding pin hit with a hammer to start the engine. The ensuing explosion forced the piston down the cylinder and started the engine.

Apart from those described, the only other manufacturer of farm tractors in the UK was David Brown, whose Cropmaster model had the advantage of being the only one with a dual seat, very convenient for those courting! I don't remember seeing a DB near home; the only other tractors in the area that I recall were a very small number of American imports. George's father-in-law had a Case; next door to my home farm there was a two-cylinder John Deere with horizontal cylinders like the Field Marshall, and on one occasion a big yellow MM (Minneapolis Moline) turned up with the threshing machine.

Although present-day tractors are often huge compared with the ones that I drove, their four-wheel drive, power steering, multiple gear ratios and sophisticated, electronically controlled hydraulics and engine management systems make them easier to drive and many times more efficient. What those of us who struggled with those early models probably envy most is the comfort enjoyed by today's drivers. We had no sound-insulated warm cabs with heaters, radios and, above all, no padded well-sprung ergonomically designed seats. The seats on all the tractors that I drove were all shaped steel bucket seats with large holes in them to allow rainwater to drain away. They were very similar to implement seats except that they were mounted on a springy steel bar to allow some of the jolts to be absorbed. Nevertheless they were not comfortable. The drainage holes tended to rub and cause sores if the driver forgot to take a sack to cover them. The footplate placement low down on the Standard Fordson allowed the driver to stand to drive it and I spent many hours operating it like this. There were no cabs and whatever the weather we continued, often half-frozen or wet through and deafened by engine noise. I wonder how many bad backs can be attributed to long hours spent on those early tractors. Nor were these early tractors fitted with safety frames or cabs to protect the driver if the tractor rolled over on a steep side slope.

Chapter 6

Autumn

THERE wasn't the same frenetic activity on farms in the autumns of the 1940s and early 1950s in North Devon as there is today. Fifty years on farmers work all hours to sow their winter corn crops before the deadline date beyond which it has been shown that there is a predictable loss of yield, whilst simultaneously harvesting root crops such as potatoes and the first sugar beet. Certainly winter wheat had to be sown but, as the effect of late sowing was not so clearly understood, there wasn't quite the rush there is today. Winter barleys were only just being developed and so we sowed both barley and oats in the spring.

Sugar beet wasn't grown in North Devon. The soil and terrain there are not suitable for large-scale production and as a consequence there is no factory to process it. However, there was still work to be done in early autumn. There were potatoes to be harvested, mangolds and swedes took the place of sugar beet, and before winter wheat was sown the opportunity to cultivate several times to kill weeds would be taken if time and weather permitted. In the absence of chemical herbicides, cultivation and rotation were the only means of controlling weeds.

Another activity in early autumn was picking top fruit. The principal top fruit grown was apples but there were plums and pears to be picked as well. Both my farms had orchards, but the largest was at my home farm where numerous varieties of apples were grown and there were a few plum trees.

I wish that I could recall all the varieties of apple, many of which I suspect have disappeared. I remember the small, dark red Devon Quarrenden, a very early apple that we used to pick and eat straight off the tree; the Gillyflower, which I suspect was the Cornish Gillflower; Tom Putt, a lovely apple that is said to have originated at Gittisham in East Devon, close to where my family originated; Worcester Pearmain; a Russet whose pre-name I cannot recall, and Bramleys, both crimson and green. There were also numerous cider apple trees but, like many farms in the area, these trees were not picked as the apples were small and sour and unsuitable for cooking or eating raw.

When I started work, the cider press and circular wooden vats in a barn next to the orchard were still in good condition, although they had not been used for some years. A few years later I helped to remove them when the space was required for another purpose. The press stood about 10 ft (3 m) tall and appeared to be very old as the threads on the wooden screw and in the wooden block that it passed through had been cut by hand.

The cider was the notorious Devon 'scrumpy', or rough cider. Whether what is sold as rough cider to the holidaymakers who stay in North Devon today measures up to what used to be made is questionable. If only a few of the tales that surrounded its making were true, hygiene controls would prevent it from being marketed today. It was always joked that the reason the vats were circular was so that the rats would run round and round until they became so intoxicated by the fumes that they would fall in and drown. This was said to enhance the flavour of the cider!

Certainly the locally brewed cider was potent, although over-indulgence tended to lead to sleep rather than to the aggressive and belligerent behaviour that so often seem to result from drinking large quantities of lager these days.

I cycled to my second farm every day, and the last mile and a half were uphill along a narrow twisty lane from the village. A neighbour of my employer regularly used to cycle down to the village pub where he was renowned for the amount of cider that he could drink. Many mornings I found his bicycle in the ditch where he had left it the previous night when he had fallen off or couldn't manage to ride it any further. I always knew how good a night he had enjoyed by how far from the village I found the bike.

At my first farm, early varieties of apples were picked as they became ripe and were mostly sold on the milk round. Most of the main crop that formed the bulk of the trees to be picked consisted of varieties that would keep, and these were stored. They were sold later in the winter or even in the spring when they commanded better prices. There was little competition from overseas in those days and the housewife appreciated the succession of English varieties, with their different qualities of crispness and sweetness, throughout the winter and well into the spring.

The trees in the orchard were old, probably planted well back in the nineteenth century, and with little regular pruning had grown so high and wide that most of the fruit had to be picked from ladders.

We used the long 24-rung rick ladders for this. Substantial one-piece ladders made of wood, they were very heavy to move about and so two people were required to raise them upright. To do this, one worker stood on the bottom rung, the second picked up the top rung and walked towards his mate pushing up the ladder hand over hand, one or two rungs at a time, until it was vertical. Once this had been accomplished it was possible for one person to carry the ladder to the tree. The ladder was held tightly at right angles to the body, with the rungs gripped low down, before it was lifted just clear of the ground. The carrier then moved off very slowly, keeping the ladder as near to vertical as possible. This was quite difficult if it was windy or if the top became entangled with a tree branch as more than two thirds of the ladder was above one's head. At the first sign of any wavering the bottom of the ladder was grounded to steady it and to enable control to be regained. If it was allowed to sway too far from the vertical it was impossible to recover or hold, and down it came. When first entrusted with one of these ladders I was warned to let it go if this happened, because the ladder was so long that when the top hit the ground it would whip and could cause a severe internal strain or rupture if I was still hanging on to it.

The ladder was laid gently onto the foliage of the tree in an attempt to avoid dislodging any apples. If any did come down there was a good chance that they would land on your head. Climbing the ladder was quite nerve-racking at first because, as I did so, the branches supporting it bent back under my weight and then suddenly flipped around the ladder causing it to lurch further into the canopy of the tree. Sometimes, this resulted in sideways twisting movements as well, but I soon realised that whatever happened the foliage was so dense that the ladder couldn't fall out of the tree and so as long as I hung on I wouldn't fall. George's father encouraged and reassured me by saying repeatedly, 'You can't vall, there be nort olding ee.' My own philosophy was based more on that of sailors on the big windjammers – one hand for oneself and one for the ship, or in this case the apples!

The apples were picked and carefully slipped into a small hessian sack supported by a rope around my neck so that the sack rested on my chest. The rope was tied to the top and bottom corners of the sack, a small apple trapped in the bottom corner preventing the rope from slipping off. Each apple was picked individually by rotating it on its stem which would twist off easily if the apple was ripe. If it wasn't quite ready a slight additional upward kink of the stem would usually be sufficient to detach it. When two fruits grew side by side on the same stem, as they often did, they had to be detached simultaneously or one invariably fell. Unless the two apples were small enough to be gripped in one hand and twisted off simultaneously, caution had to be thrown to the winds and both hands had to be used – leaving none to hold the ladder!

As the bag filled up it bulged out in front of me and

began to push me away from the ladder. This made it difficult to reach out for apples that were further away without the bag swinging about and pulling me off the ladder, and so I would have to descend. Back on the ground, the apples were gently transferred to wooden-chip baskets and these were carried, one in each hand, out of the orchard, across the yard and up the back stairs of the farmhouse to the room where they were to be stored for the winter. The walls were lined with waist-high shelves. These had a vertical lip at the front that prevented the apples from rolling off the shelf and enabled them to be piled up against the wall in a heap. This was not in accordance with the usual recommendation in gardening literature that each apple should be individually wrapped and positioned so that it does not touch any of its neighbours. Nevertheless, the majority of the apples that were stored seemed to keep perfectly well.

I found apple picking quite tiring. Apart from constantly stretching and reaching and the weight of the bag swinging from the neck, my legs ached from standing for long periods with the rung of the ladder wedged into my insteps.

Although we had no sugar beet, we did have other roots to harvest. The most important was mangolds, or mangels or mangel-wurzles, as they are variously called. They were used to feed cattle, sheep and even horses. They are seldom seen these days as other crops that require less labour and provide better-quality nutrition for the animals have taken their place. However, they were very important for livestock right through the first half of the twentieth century and they were grown on both my farms. The varieties that we grew were orange and yellow in colour, globular in shape and variable in size, the largest the size of a football.

The roots were pulled by hand and the tops chopped off with a long knife or twisted off. We usually cut them off, although twisting was recommended as cutting into the crown could cause the root to bleed and rot more quickly. Cutting was more hazardous too, particularly late on a cold and wet autumnal afternoon as it became dark, when it was easy to become careless. After they had been pulled, four rows were usually thrown together into a single row so that when they were to be loaded there was room for a cart to be driven between the rows. We loaded them, one or two at a time, by scooping them up with a long-handled muck fork with its four curved prongs, turning the fork over to flick them off into the cart. Forking them required more skill but was much easier and less tiring than continually bending to pick up the roots and then throwing them into the cart, although as with topping it was argued that piercing the roots with the fork left holes where rot was likely to start.

A horse and cart made loading easy. There was no need to lead the horse down the row of roots as it would walk forward a few paces in response to voice commands, enabling loading to continue uninterrupted. The cart used was referred to as a butt. It differed from the cart used in the harvest field as it had a solid front and sides, and a tailboard with half hinges at the bottom so that it could be dropped off out of the way when lowered. A big advantage of a butt was that it could be tipped to empty the load and the horse could help to do the tipping. Like the hay cart the butt had shafts for a single horse. In the case of the butt these were hinged halfway between the central axle and the front of the cart body, so that when the shafts were moved back they provided leverage and a turning moment that lifted the front of the cart body. In practice, the driver pushed up the front of the body with one arm to help to start the butt to tip, urging the horse to 'cum back' and simultaneously pulling back on the reins held in the other hand. This induced the horse to edge back and use its strength to raise the front of the butt and tip the load.

Once the butt was loaded the horse was led back to the farm, but on the return journey to the field one could ride. This meant perching on the narrow edge of the board forming the front of the cart body cushioned only with a sack, both feet placed one in front of the other on one of the shafts to maintain balance as the wooden-spoked wheels jolted over the uneven ground. Perhaps, after all, second-class riding might not have been better than first-class walking!

There was still some discussion locally about using tractors and trailers for this type of work, even after rubber tyres had replaced the steel wheels fitted with spade lugs that were not legally permitted on the road without road bands. Every time that a length of row had been cleared, the tractor had to be moved forward and someone had to stop loading, climb onto the tractor, start it, move it on, stop it and get off again. Although the tractor trailer would carry more and was quicker on the road, many of our fields were close to the farm so this was a limited advantage. However, the biggest disadvantage was that the first trailers that we had could not be tipped and had to be unloaded by hand. This was true of most farm trailers until hydraulic tipping systems became available, although a few were equipped with vertical screw tipping gear operated by a hand crank rather like the tipping gear on lorries of the period. The horse and cart continued to keep its place in the small fields and narrow lanes of North Devon for some years, but by the time I finally left the farm in the mid-1950s the tractor was very rapidly displacing the horse from most farms for good.

Mangolds are susceptible to frosts and so the crop had to be lifted before these were likely to occur. A few loads for immediate use were dumped in a building in the farmyard, but most of the crop was stored in a clamp in a shallow depression next to a bank in the rickyard. The clamp was covered in straw and earth to keep out the frost. When the mangolds in the barn had all been fed to stock, several loads would be transferred from the clamp.

The mangold tops were not wasted. They were either forked up into carts and taken back to the farm to be fed to cattle, or sheep were folded on the field to eat the tops in-situ. To prevent the sheep from roaming the whole field and spoiling more than they ate, they were confined in a small area enclosed by a row of hurdles across one end of the field. Once they had eaten most of the tops in that part, the hurdles were moved to enclose a larger area and this procedure was repeated until the sheep had cleared the whole field. Apart from the feeding value of the tops at a time of year when there was little grass growth, even in a warm damp climate like that of North Devon, folding the sheep across the field ensured that they covered it uniformly with their droppings so providing manure for next year's crop.

Mangolds had to be sliced before they were fed because animals could choke on roots that they attempted to swallow whole. Although machines were available to do this, we chopped the roots by hand on a chopping bench. One end of the long chopping knife was pivoted on a swivel mounted on the bench. The handle at the other end would be worked up and down and swivelled from side to side to slice the beet and then to sweep them into a galvanised bath. This was carried to the stock, and the slices tipped into the troughs or mangers where they were quickly eaten by the animals who clearly enjoyed them.

On my second farm we also grew swedes. They were grown as a 'catch crop' after an early crop had been harvested. Swedes were something of a gamble. Some years there was a good market for them in London, presumably when other vegetable crops were in short supply. In other years, when there was no such market, the sheep were folded on them behind hurdles in the same way that mangold tops were fed. Swedes were a good food source for the animals as they have a higher dry matter content than mangolds.

On both my farms we also grew cow cabbage, or *flat polls* as we called them. These were very large and coarse drumhead cabbages, flattened rather than spherical in shape and sometimes 2 ft (60 cm) across. They weighed up to 70 lb (30 kg). They were cut with a long knife and forked up into the butt one at a time to be transported back to the farm. They were primarily grown for the cattle to eat.

In the year I spent at my second farm, cow cabbage for transplanting was the early crop that preceded swedes. The seed was sown in the spring but the seedling plants were pulled in early June and sold on to other farmers to transplant and grow on in their fields. We pulled them by hand and tied them in bundles of twenty or more. Each farmer would buy several thousand plants and so large numbers had to be pulled. We pulled them by the handful and with practice could judge when we had enough for a bundle.

We also transplanted the cabbage that we needed for our own cattle. The plants were spaced out on the ground beside drills marked out with the ridging bodies normally used for potatoes. We walked along the row picking up a plant with one hand and slipping it into a slot made in the ground with a short-handled digger held in the other. The blade of the digger was driven into the soil with one blow and pulled back to create the slot. This was pressed shut with the toe of the boot firming the soil around the plant as we moved on. Once the rhythm had been established, the whole process was done at almost walking pace. It may sound rough and ready but my employer used to say, 'Ang 'em in the ground and they'll grow.' He was right, too, because the success rate was high. No doubt the warm and humid weather in North Devon was a major factor in the successful 'take'.

At my second farm, marrow stem kale was grown in addition to cow cabbage. The stems grew waist high, very thick and close together to give a high yield. Kale was grown for the dairy herd to graze late in the year when there was no grass left. The area of kale made available for the cows to eat each day was controlled in the same way as the mangolds and swedes for the sheep. However, instead of hurdles, an electrified fence was used to surround the area to be grazed.

The electric fence consisted of a single wire strung on insulators nailed to a row of posts stretching across the field and connected to a fencer unit powered by a rechargeable battery. When a cow touched the wire, it completed the circuit to earth and received a shock that made it recoil from the fence. Although the fencer unit stepped up the battery voltage to several thousand volts by means of a transformer and a simple 'make and break' switching system, the current was minute and the animals came to no harm. I frequently received a shock by accidentally touching the wire after switching on the unit, or when testing the system by the crude but effective method of holding a grass stem against the wire. This was supposed to lessen the shock by increasing the resistance but didn't always do so if the stem was wet or contained too much sap. The shock would tingle and jerk my arm off the wire, as though I was touching a spark plug on a running engine. Although erecting an electric fence is quicker and cheaper than using hurdles, it hadn't at that time been adapted for farm livestock other than cattle.

Kale wasn't the easiest of crops to electric fence as the crop was so thick that it was difficult to force a way through. After the fence had been erected, all the kale stems near it had to be removed so that there was no chance of a stem falling against the fence or being pushed against it by a cow and short circuiting the wire to the ground. This was not the most pleasant of jobs on a day when the waist-high kale was wet with rain, or covered with frost after a cold night. I wore long waterproof leggings attached to my belt to keep my legs dry, but my hands were often blue with cold by the time the job was done.

The cows soon learnt not to touch the fence. However, they would go down onto their front knees so that they could stretch under the fence without touching it to reach kale stems on the other side. It was not unknown for a dominant animal to drive another against the fence deliberately in order to push it down onto the ground and short circuit the wire. The animals could then get at fresh kale, rather then clear up the allocated area. From time to time, when we went back to collect the herd for afternoon milking, we would find the fence down, the cows scattered all over the field and a lot of kale trampled down.

Modern electric fencing posts are slender steel rods with an insulated pigtail top that can be pushed into the ground by hand. Our posts were wooden stakes that had to be driven in with the back of an axe. I always carried an axe and a few insulators and nails with me when I went out to move the fence so that, if I was short of a couple of posts, I could cut them out of the hedge and use the back of the axe to nail on the insulator.

Throughout the autumn, when time and soil conditions allowed, the corn *arrishes,* or stubbles, were cultivated and then ploughed. When I started the stubbles were still being raked with the horse hay rake after the sheaves had been cleared and the rakings carted home loose to be fed to the poultry.

Cultivations to control weeds started immediately after the corn crops had been cleared to encourage weed seeds to germinate so that they could be killed by further cultivations. This was done with a cultivator fitted with a number of 'tines', vertical legs with A-shaped blades at the foot, which disturbed the soil and undercut the weeds as it was pulled across the field by the tractor or, before it arrived, by horses. We called this operation 'scuffling' and the cultivator a 'scuffler'. If time allowed the land was scuffled again to kill the new flush of weeds that had been encouraged to germinate by the first cultivation. Ultimately, the land was ploughed and the final growth of weed turned in under the furrow so that it would not regrow.

The time available to do this was limited if autumn-sown winter wheat was to be grown. This was drilled in October, if possible. I doubt if many of my farmer friends then were aware of the research into the effects of later drilling, but sowed the seed in accordance with traditional dates established by trial and error.

Ideally, all the remaining ploughing was completed before Christmas and the threat of serious frost, although crops such as swede and kale which stayed in the ground until late in the year prevented some land from being ploughed until the spring.

Ploughing leaves the surface furrowed, exposing the maximum surface area to be weathered during the winter. Continual wetting and drying, and the effects of frost during the winter months tend to break up the surface layers, reducing the time and effort required to produce a seedbed in the spring. This is of less consequence today when most arable farms have far more tractor power than is really necessary, and power-operated cultivation machinery that can force a tilth except when the ground is too wet to work on. Our small tractors and simple cultivators and harrows, almost all of which had been developed from horse-drawn predecessors, did not have the same effectiveness, and we had to rely on the help of the elements and only cultivate when conditions were favourable. There were occasions at my first farm, in very dry conditions, when cultivations resulted in small clods that resisted all our attempts to break them down to a fine tilth with the cultivation implements at our disposal. The clods rolled away from tined implements rather than being shattered by them. Our remedy was to use a heavy roller that crushed some clods and compressed the remainder into the soil surface, preventing them from moving so that subsequent passes with a cultivator shattered them.

My first close-up experience of ploughing was gained from riding on the tractor soon after I started spending time at the farm near my home. George did almost all the tractor work on the farm; he was the only one with a licence at the time. I used to spend hours standing beside him when he ploughed or did any tractor work.

The Standard Fordson on the farm was an ideal tractor for this as there was a cramped space to the left of the driver where a passenger could stand. This was an area of flooring at a low level just behind the rear axle so close to the ground that a passenger could step on or off with ease, as the tractor only moved at walking pace when doing field work. One could also slip off, as I discovered one cold and wet afternoon. I had got cold and stiff and as the floor plates were wet I suddenly slithered off and my foot became trapped under the drawbar of the plough. Fortunately, George saw me disappear and stopped the tractor before the wheel ran over me. The other regular passenger on the tractor was a fox terrier who used to ride for hours lying on a sack on the axle housing, occasionally jumping off to search the hedgerows for a time before climbing back on again.

I learned all the basic skills and techniques of ploughing while riding with George. The Ransomes plough trailed behind the Fordson was carried on three wheels, one running in the furrow, one up on the unploughed land and a castor wheel at the back also running in the furrow. The plough was fitted with three plough bodies and so could turn three furrows simultaneously. There were three principal controls accessible from the tractor seat, and I was soon operating them at George's instruction. Two of the controls were crank handles that were turned to control the depth of ploughing and to level the plough so that all three bodies turned equal furrows. The third control, a lever, altered the front furrow width and was used to remove any kinks that developed in otherwise straight furrows. Farmers and their workers took great pride in their ploughing and in producing dead straight and even furrows.

At the end of each furrow the plough was lifted out of work by pulling a cord that engaged a clutch on the wheel running on the unploughed land so that as the wheel turned it raised the plough. Then the tractor and plough could be turned around on the headland to face the other way ready to start the next run down the field. However, if the ground was wet and the surface greasy the land wheel tended to slip and the plough stayed in the soil. This meant the headland would be ploughed and prevent the turn from being made. Cleats bolted to the wheel rim helped to prevent this but it took time to fit them and to remove them before towing the plough on the road.

Wheel slip also affected the tractor. Although its wheels were fitted with spuds – the triangular-shaped lugs that dug into the ground – they could still slip in very damp patches and if very wet the tractor would stop, although the wheels would still be turning. Within a couple of revolutions, if the clutch was not pushed down immediately, the wheels dug themselves in too deep to extricate. We always carried a long heavy chain with a ring at one end and a hook at the other on the tractor. As soon as the tractor began to slip and before it had dug in, the plough was unhitched, the tractor driven forward onto drier ground and the plough towed out with the chain. Modern tractors have differential locks, independent brakes and often four-wheel drive, all of which make wheel slip less of a problem unless the conditions are exceptionally severe.

In addition to the three adjustments made from the tractor seat, there were other alterations which had to be made with a spanner when the tractor and plough were stationary. For commercial ploughing, one of the most important was the setting of the skim coulters. These looked like miniature plough bodies designed to slice a corner off each furrow so that when the furrow was turned it rested tightly against its neighbour. This sealed any gap between furrows through which weeds or grasses, that had been inverted and buried when the furrow was turned, might grow, thus ensuring that they would die. Thin lines of green shoots along each furrow, or worse still lumps of weed and turf strewn over an area of ploughing, are indications of poor plough setting, the coulter settings likely to have been the major cause.

George also showed me how to set out the ploughing. First of all a 'headland' that provided room for the outfit to be turned was marked out at each end of the field with a shallow furrow parallel to the hedge. The line to be followed was determined by pacing out from the hedge at intervals and planting markers. An alternative method was for me to walk along close to the hedge holding one end of a string of the right length, with the other end tied to the radiator cap, while George drove the tractor keeping the string taut.

Work commenced by ploughing down the field and back, turning two sets of furrows against each other to form a straight ridge down the length of the field, stopping short at the headland mark at each end. George ploughed around and around these furrows until the strip of ploughing, the 'land', was 15 to 20 paces wide. Wider than this, it took too long travelling along the headlands from one side of the ploughed land to the other, and so a new ridge was set out further along the field. Then the ploughing continued around the new ridge. The position of the new ridge was set out parallel to the completed ploughing by measuring across from the first ridge at intervals along the field and pushing markers into the ground. Two markers close together were required at the far end so that by keeping them in line a dead-straight furrow could be drawn.

When I started driving I discovered that it is almost impossible to drive in a straight line to a single marker. Whenever I attempted to do so I saw with horror a beautiful curve behind me when I looked back, instead of the required straight furrow. Straight ploughing allowed quick clean finishes to be made. However, when cultivating or doing other tractor work, reasonably straight lines across the field were still desirable as they made it easier to match previous work without wasting time overlapping or doing unnecessary short work. Reasonably straight passes across a field could be achieved by lining up the two fuel tank caps and the thermometer on the radiator on the early Fordsons with a prominent feature at the hedgerow at the opposite side of the field, such as a tree or fence post.

Finally, a finish was made between the two lands, leaving a narrow, shallow furrow where they met. It was much easier to make a clean, even finish if the furrows at each side of the last run were straight, parallel and without kinks. The skill was to keep the land surface as level as possible by avoiding deep finishes or high starting ridges. George usually tried to remove any unevenness from the previous year's ploughing by setting the ridges in last year's finishes.

The remainder of the field was ploughed in the same way until only the headlands were left unploughed. The fields were seldom truly rectangular, and so a few short runs often had to be made to plough the triangles or odd shapes left as the ploughing approached the side of the field. I learned later that our system for working our way across a field was not totally in line with the method recommended in textbooks, but it suited our small fields.

Finally, the headlands were completed by ploughing round and round the field. The furrows were turned inwards towards the centre of the field one year, and outwards the next to prevent either a deep furrow or a mound being created at the edge of the field. The plough was offset for the very last circuit around the field by swinging the towbar on the tractor to the side so that it could plough right into the foot of the hedge, leaving nothing unploughed. When the work was finished there was just a tiny area of unploughed land in each corner of the field that the plough could not reach. Not many years previously, a worker would have been sent out to dig the corners by hand and plant cabbages there.

Most of my experience of ploughing on my own was gained when I worked full time at my second farm. The Fordson Major tractor there was equipped with a two-furrow reversible mounted plough also manufactured by Ransomes. This simplified the ploughing operation considerably. As the plough was mounted on the hydraulic lift system fitted to the rear of the tractor it could be raised with ease and certainty at the end of the furrow, and could be reversed into the corners of a field to ensure that almost everywhere was ploughed. Being reversible it had two sets of plough bodies, one set designed to turn furrows to the left and the other to the right. One set was in work while the other was balanced upside down above it. At the end of the field a U-turn was made and the positions of the two sets of bodies were changed over so that the plough bodies now in work turned furrows the same way as the previous ones. This meant that ploughing could start at one side of the field and continue without a break to the other, leaving only the headlands at each end to be completed. This avoided setting out lands and all the marking out, starts and finishes associated with them, although the headlands still needed to be marked out before starting. The furrows were all equal in size and parallel but not necessarily straight because they followed the line of the hedge which was seldom straight.

The chief penalty of using the reversible plough was that it was limited to two furrows because of the extra weight of the second set of bodies. This disadvantage was offset by the time saved in setting out and running along the headlands compared with using a trailed plough and undoubtedly less skill was required to achieve a good result.

I discovered an unexpected snag when using a reversible plough that no one had warned me about. The plough bodies were turned over automatically at the end of the furrow as the tractor was turned around. This was fine until I started to plough a field with a pylon in the middle. When I reached the pylon I lifted the plough out of work, drove around the pylon and dropped the plough back into work. I forgot that lifting the plough automatically reversed the bodies, and when I looked back I discovered that as a result the furrows were being turned the wrong way and that a big gap next to the ploughed land had been left unploughed. Unfortunately, it is very difficult to correct this type of error without making an even greater mess. I cannot believe that my employer did not notice the mess and guess what I had done but he never commented. I soon found out how to override the automatic mechanism that reversed the bodies and so prevent this accident happening again.

Although I was given very little opportunity to plough at my home farm, as it was George's favourite tractor job, I was harrowing with the tractor by the time I was 13. Current safety regulations do not permit youngsters even to ride on a tractor until they are that age. At first I steered the tractor while George stood in my usual place beside the seat, but within the year he would leave me to it and go off and do something else in the same field within sight of me.

There was really little to do on a Standard Fordson when cultivating except steer. All cultivating was done in one gear, and if I needed to stop all I had to do was to depress the combined clutch and brake pedal and clip the hook over it to hold it there.

Turning when the headland was reached was the only real difficulty. The draught force from the cultivator being pulled tended to make the tractor continue in a straight line, and also had the effect of transferring weight away from the front wheels. As a result, when the steering wheel was turned the tractor responded very slowly and the front wheels scrabbled along sideways until they obtained some grip. It was often the foot of the hedge bank that finally pushed them around.

Although field work carried on regardless of weather, provided that the soil wasn't too wet, none of the tractors of the time had any weather protection or heaters, so we worked on in cold, wind or rain muffled up in thick coats and hats. Ex-services greatcoats were the fashion of the time. George had been a member of the Home Guard and wore his long khaki greatcoat, and when I grew older and taller I appropriated my father's black ARP coat. These garments were long and warm but thick and heavy, particularly when soaked with rain. Like everyone else I wore a cap all the year round. This kept the head warm and dry and hair clean. In summer it gave protection from the sun's heat and glare. For most of the winter we wore wellington boots. In drier times during the rest of the year, I, like the others, wore hobnailed boots with short leather gaiters above them.

Chapter 7

Winter

ONLY the sheep were out in the fields during the winter. All the cattle and the horses were housed indoors for several months. Winters in the river valleys in North Devon are warm but very wet and as a result grass usually grows almost right through the season. However, because it is so wet the ground becomes badly *paunched*, that is 'poached', if the cattle are left out, particularly near gateways and feeding areas. Their hooves churn up the soil, creating large areas of mud with no visible grass. These areas are difficult to reclaim in the spring, and there is a long wait before the grass regenerates and can be grazed.

When the cows were being prepared for milking their udders and teats were washed with a cloth and warm water. If the cows had lain down in the mud their udders and teats became filthy and took ages to clean, particularly if the mud had dried on.

The sheep were left out as there was no room for them in the buildings, although they were given supplementary feeding of hay and roots. They were moved into a field close to the farm buildings in late winter at lambing time so that they could be regularly checked and given help if it was required. In later years, after I had left the area, George, like many others, built special accommodation for the sheep in which they could overwinter and have their lambs. This reduced the mortality amongst young lambs and made it much easier to look after the flock. It was also much more pleasant for the shepherd.

Throughout most of the winter the dairy cows were kept in and did not leave the cowshed where they were milked. Apart from milking, the routine for looking after the cows included feeding them twice a day with hay and, probably, chopped mangolds. Cleaning out twice a day was the other routine task.

The young cattle and the fattening bullocks were kept in open-fronted buildings called 'linhays' that opened at the front into enclosed yards where they could obtain water and be fed. They were not cleaned out during the winter. More straw was put into the yard regularly for them to lie on. In the spring when the animals were turned out the accumulated muck was loaded into carts and taken out to the fields. The very young calves were kept in pens in a barn in the warm throughout the winter.

Much of the short winter days were spent looking after the stock, as the horses were also in their stables during the winter and food had to be taken to the sheep in the field. There was little field work to be done. All the mangolds and other root crops had been lifted and carted back to the farm and any winter ploughing completed before Christmas. However, there was always plenty of hedging and ditching to be done.

Small irregular fields bounded by hedges grown on tall banks and sunken winding lanes are characteristic of much of the North Devon countryside and contribute greatly to the beauty of the landscape. This landscape is now thought to be very ancient, possibly iron age, and was formed by early farmers who cleared the land from the wild by stripping off the turf. This was stacked around the small areas that they had laboriously cleared to fence them, both to denote their ownership and to protect the crops from wandering animals. The network of narrow lanes left to provide access to the fields and to allow people to travel from place to place became rutted in the course of time because of constant use by animals, farm carts and travellers on foot. On the slopes in the wet climate these were washed out, aided no doubt by farmers who shovelled up the rich mud to spread on their land as manure, giving rise to the sunken lanes common to the area.

It is a very curious contradiction that in the midst of this ancient landscape Braunton Big Field still exists as one of the few examples of open-field farming left in England. Systems of communal farming in huge unfenced fields had evolved by the end of the ninth century, although they were not adopted until later in some areas. These are often described as the Three Field system, inaccurately as there were often different numbers of fields associated with a village as well as common land used for grazing. This system of communal farming predominated in much of England in an unbroken area stretching from the South coast to Northumberland but excluding Wales, the Welsh borders, Devon and Cornwall and parts of East Anglia, and persisted in some areas until the last major enclosure movement in the eighteenth century. It is believed that farmers used to work a number of strips in each field so that both the good and bad land were shared out evenly amongst them, but there had to be agreement on the crop to be grown in each field. A number of farmers still have rights on Braunton Big Field, although many of the strips have been consolidated. It seems unlikely that many of the rest of the small irregular fields in North Devon were ever farmed in this way. However, the arrangement of fields around Combe Martin has led to suggestions that some form of communal farming took place there also and, therefore, perhaps also existed elsewhere in North Devon. If so, it must have been organised within the small fields that were already in existence and been enclosed at a much earlier period of history.

In the mild wet climate the hedges grew tall and thick. They soon reached the stage where they shaded large parts of the small fields. Such areas lay damp and cold in the spring so that crops were slow to germinate and mature, and grain ripened unevenly and was slow to dry in the stook.

Laying the hedges and maintaining the banks was a slow business and we only managed one big hedge each winter. The intention was to work right round all the hedges on the farm in rotation before any of them became too high and the saplings too thick to lay. Unfortunately, the work had got behind because of the lack of labour during and immediately after the war, and many of the hedges had got out of hand. This was true of most of the farms in the area and the situation did not improve until mechanical hedging machines became available. Unfortunately, these machines only cut the hedge that is still present and do not repair and prevent gaps as does laying a hedge, although regular cutting does encourage the hedge to thicken.

The objective when laying a hedge, apart from reducing its height, was to lay down young saplings horizontally on top of the bank to form a stock-proof fence that would grow vigorously to form a thick new hedge.

Hedge laying was done in winter when the sap was down. Suitable tall but thin saplings were selected for laying, and the remainder of the heavier growth and brushwood was cut out with an axe and a 'browsing' hook. This was short and stout with a slight curve. Occasionally old, thick *moats*, stumps that had usually been pollarded several times, had to be sawn off with a cross-cut saw as they were often a couple of feet thick. This was a slow hard job, with one sawyer balanced awkwardly on the top of the bank and the other on the ground being smothered in sawdust. The saw was about 6 ft (2 m) long with large triangular teeth set very wide, and it had a straight, vertical wooden handle at each end.

Once the bank top had been cleared, the saplings selected and retained for laying were partially cut through with the browsing hook, trimmed and pulled down onto the top of the bank. There they were secured with 'crooks' driven into the bank with the axe head. The crooks were cut with the browsing hook from forked lengths selected from the discarded hedge timber. The saplings were laid side by side and on top of each other, overlapping along the full length of the bank, to form the basis of the new hedge. These on their own would prevent a sheep climbing over, particularly as the Devon Closewools that we kept are not noted as climbers or jumpers. Once spring arrived the side branches of the saplings that had been laid would soon start to grow straight up to form a thick impenetrable hedge that would not need attention for many years.

Once laying was complete the thicker timber cut out of the hedge was trimmed of its side branches and carted home to be sawn up with a circular saw for fence posts or for logs. Some of the trimmings were bundled together and tied with a withy to form a *faggot*, colloquially a 'vaggot of 'ood', which was burnt on the big open fireplaces in the farmhouse. Traditionally, ash faggots were burnt at Christmas. The remainder of the trimmings were scraped together and burnt in the field, or if it was a grass field they would be carted to an arable field where they could be burnt without damaging the grass sward. Tending a big blazing bonfire was a very pleasant occupation on a cold winter's day.

When a large pile of timber cut from the hedges had accumulated in the inner farmyard it was sawn up on a portable circular saw bench that was positioned close to the heap. The saw blade was driven from the tractor pulley by an endless belt. It was inconvenient and, more important, unsafe in the case of an emergency to have to walk backwards and forwards to the tractor every time that it was necessary to start or stop the saw. To avoid this the saw bench was equipped with a 'loose' pulley next to the driven pulley, so that the belt could be shifted over onto this pulley when it was necessary to stop the saw. The belt was moved over from one pulley to the other by a fork spanning the belt that pushed it across when the control lever was operated. However, the saw blade was 3 ft (1 m) or more in diameter and possessed considerable momentum when rotating at full speed, and so there was a delay before it came to a halt.

Fence posts and stakes were cut and pointed from the bigger timber. This was halved or quartered, or squared lengthways depending on its size. If the wood was unsuitable for this purpose it was sawn into logs. George did the sawing and I passed the branches to him. The big moats sawn from the hedgerows required both of us to lift them onto the saw bench and they were so cumbersome and awkward to feed onto the saw blade that it was very easy to jam the blade and stop it. The belt slipped when this happened and in doing so acted as a safety device.

Before sawing we had to examine all the timber, the moats in particular, for nails or embedded barbed wire which would notch the saw blade. These were often hard to spot because the bark had grown over them. To sharpen the blade and remove any notches the blade was clamped in situ in the bench with wooden wedges to stop it rotating. Each tooth was then sharpened with a triangular file, with each one given the same number of strokes so that the balance of the saw blade was retained, and the profile of each tooth retained. Rotating an out-of-balance blade, particularly a large diameter blade, at speed can cause serious and dangerous vibrations to be set up.

Sawing was a particularly hazardous job. We always took great care in walking around on the rough, sloping ground near the saw bench to avoid slipping or tripping, particularly when carrying heavy or awkwardly shaped timber and when manoeuvring the wood on the saw bench. George was very careful to keep his hands well to each side of the blade as he pushed the lengths of timber to be cut towards it, and to use long lengths of wood to push cut pieces and rubbish clear of the blade.

The banks were maintained at the same time as the hedge was laid. During the years since the work had last been done, rain had washed soil down from the top of the bank leaving depressions and gaps over which stock would ultimately be able to scramble. These were made good by 'casting' – digging out turves at the base of the bank where the eroded soil had lodged and using them to build up the top of the bank and to fill in any holes. The turves were dug out with a 'tubill', a device similar to a pickaxe but with two wide, flat blades set at right angles to each other instead of points. We used the Devon shovel, with its triangular blade and long, curved shaft, to place the turves on the bank. The final result when well done was a near vertical bank of uniform height, free of gaps, which farm animals would find difficult to scale.

Any ditch at the base of a hedge was cleaned at the same time. All the silt and weed that had accumulated was shovelled out. Any solid material was added to the bank and the rest was spread on the field. The aim was to provide a clear way for water to trickle steadily along the ditch, usually following the general slope of the field and without forming ponds at any point. We worked backwards along the ditch, starting at the top end so that the water flowing down towards us indicated whether we were maintaining the correct gradient. This was a wellington boot job, particularly if there was an accumulation of water in the ditch. As we were usually standing in squelching mud it was difficult to extricate my foot when it was necessary to move and very easy to step out of the boot and leave it behind.

Stone walls are not common in the valleys of the Taw and Torridge, but we often built dry-stone ends to the bank where it was broken by a gateway. All the gates were traditional five-bar wooden gates with a diagonal fixed from one top corner to the opposite bottom corner to prevent it sagging. They were quite narrow, only wide enough for a cart to pass through, and most of them had to be widened or made into double gates when the combine and wider implements came into use. The gates were hung on substantial wooden gate posts, usually oak for long life, although we had a number of stone ones as well that were probably very old.

From time to time we had to move or replace a gate post, either because the gateway needed to be widened or because a wooden post had rotted. Even oak posts rot ultimately. This often proved to be a time-consuming task. The biggest problem was the removal of the old post. We had no power tools or mechanical diggers and so it had to be dug out. This was particularly awkward if only the stump remained, as there was little or nothing to pull on. If the post was still whole it helped, as it could be wobbled backwards and forwards to help ease it, using its length as leverage. We never concreted around the bottoms of posts to hold them in place as moisture accumulates at the surface of the concrete where it cannot drain away and causes the post to rot at that point. We set the posts in at least 2 ft 6 in. (75 cm) deep and rammed stones around the base. Layers of soil and more stones were added, each layer being rammed in. Once set in like this they never moved, and water could seep away down through the stones.

If we had to dig a hole for a post to be set in we used a 'bar iron' to loosen the soil. This was a heavy iron bar, 1 in. (25 mm) thick and 5 ft (1.5 m) long, with a point at one end and a circular mushroom head at the other. The pointed end of this heavy bar was driven down into the hole being dug to loosen the soil and prise out any stones. Once the hole had been completed and the new post lowered into it, stones and soil were shovelled into the hole around it and were consolidated with the same bar iron, used head end downwards. The same tool was also used for inserting pointed fencing stakes, electric fence posts and hurdles. The pointed end of the bar was driven into the ground repeatedly and wobbled around to make the hole for the stake, which was then driven in by balancing the bar iron in the air and then thumping it down head first onto the stake. This was quite awkward as the weight and length of the bar made it difficult to balance the bar and keep it vertical, particularly as it was at head height when driving in a tall stake.

One other winter task that I was introduced to at my second farm was stone picking. We did no hedge laying that winter so it was probably an activity to keep me busy when there was little other field work to be done. I was sent out with a horse and cart to collect large stones from the surface of arable fields and to use them to fill in ruts in field gateways. I considered it rather a waste of time as there were few stones large enough to interfere with cultivations or damage harvesting machinery.

Towards the end of the winter the sheep lambed. Unfortunately, except at weekends while I was at school, I had little opportunity to observe this or to take part in it at either farm. University vacations did not coincide with our lambing season, and the flock at the second farm was small and managed by my employer on his own. Of course I did see lambs born and occasionally fed the 'pet' lambs, that is ones that had been rejected by their mothers or more commonly lambs that were twins or even triplets whose dam hadn't sufficient milk for all of them. The women of the farms generally looked after these pet lambs, and there were usually two or three in the kitchen, either wrapped in blankets to keep them warm or occasionally following the women around the room.

Although they were rare, some of the most memorable moments of my farming experience, indeed of my whole life, are associated with those lambing times. I often accompanied George when he went out late at night to check on the flock and to spot any *yawes*, that is ewes, about to lamb. The flock was always in a 'plat', a small field, just across the road from the farm. We would go out armed with a powerful torch and as George shone it around the field we would see the dim shapes of the sheep, their eyes shining green in the torch beam. Occasionally, a pair of red eyes would indicate the presence of a rabbit. One of our main objectives was to spot any ewe that had gone away under the hedge on her own, a sign that she was close to lambing.

The nights I remember most were the clear frosty nights when the sky was velvety black and studded with myriads of bright stars. It was unbelievably quiet. The only sounds were an occasional bleat, the hoot of the local owl and small unidentifiable animal noises. There was no traffic noise, no light pollution and there was no one else about. I recall one such night as clearly as if it were yesterday. While we were out the Albert Memorial clock in the town square struck, probably ten o'clock, and we could hear it quite clearly although we were more than two miles distant. I wished that sublime moment could last forever.

In addition to checking the sheep at night during lambing time, it was routine during the winter to walk around the farmyard each evening to check that all the animals were secure and well. The only electric light was in the dairy and to visit the stables and the cattle in the yard we had to take a lantern with us. We had a Tilley lantern, a big advantage over the earlier hurricane light, because the paraffin in the reservoir could be pressurised with a small pump fitted on the side. The result was a brilliant white light that lit up the whole yard when we went outside.

Chapter 8

Threshing

EVEN as a very small boy, long before I started to help on the farm, I can remember the excitement of the arrival of the threshing machine.

Few cars or lorries passed our gate at any time and so the sound of any approaching vehicle resulted in a dash to the front gate to see what it was and to watch it pass. The threshing machine was towed by a steam traction engine and so the chuff-chuff of the engine, the plumes of smoke and the rumble of the iron wheels announced its impending arrival long before it came into sight. Even if I didn't see it arrive, it was soon obvious that threshing was taking place from the continuous hum and the cloud of dust that enveloped the nearby rickyard, almost hiding the machines.

The traction engine passing my gate pulled a train of trailers, first the threshing machine, next a large wooden hut on four iron wheels in which a supply of coal and all the tools and gear were carried, and finally a water tank. Occasionally, a reed comber or a straw baler was also included in the train.

From 1859, when the first traction engine appeared, until 1896, a traction engine was allowed to tow as many as nine vehicles plus a water carrier. There is on record correspondence congratulating a manufacturer that his engine was used to pull ten wagons. It would appear that traffic violations are not just a recent phenomenon! Negotiating corners must have been a nightmare. The legislation in 1896 limited the number of trailers to three plus the water carrier.

Other traction engine users also took advantage of this permissive legislation. One of the events of the year that I looked forward to, when a small boy, was the arrival of the rides and amusements for the three-day Barnstaple fair that was held annually in September, as it still is. They were not allowed to arrive until the Sunday before the Wednesday opening of the fair, and so a procession of their vehicles passed that day. Many of them came from Bude, and their route took them close to where I lived. As a small boy I was taken to the top of Sticklepath Hill on the Sunday afternoon to watch them arrive. Many of the rides were still operated by beautifully maintained fairground steam engines and each one arrived towing several colourfully painted wagons and a large live-in caravan on four wheels for the family. There were a great many of them as Barnstaple fair was a very big gathering, reputed to be one of the largest in the south-west. It was an unforgettable sight.

There were still a few steam lorries operating in the area even after the war, and I can recall how slowly they crawled up that same hill going away from the town. All the road resurfacing in the area was done with the aid of a steam road roller that also arrived with its train of tool wagon and water tank. Now all these steam vehicles are only to be seen at steam fairs. Although these are enjoyable nostalgic events, the sight of the beautifully maintained engines processing around a show ring or taking part in games reduces them to interesting historic relics rather than recognising their historical importance as power sources and for haulage when there was no real alternative. To me, when a boy, they were a source of wonder and delight. To have seen them travelling and working for real is a memory that I treasure.

It shouldn't be forgotten that steam ploughing engines made it possible to plough heavy land that was virtually unworkable with horses, and enabled large tracts of heavy land to be brought into cultivation in the second half of the eighteenth century. The engines were equipped with winch drums mounted under the boiler that enabled a plough or heavy cultivator to be pulled backwards and forwards across a field by means of a wire cable. A few sets continued in work until the 1930s. A few winch engines still exist and appear at traction rallies fully restored to working condition.

Even before the war motor vehicles were scarce on the main roads of North Devon. Car ownership was confined to the wealthy and the few who needed them for their work. Almost all workers lived near their place of employment and either walked or cycled there each day or went by bus. The bus network seemed to be more user-friendly in those days and there were services that enabled the public to travel freely in the area although even then the rural areas only had services to Barnstaple on Tuesdays and Fridays – market days. During the war petrol rationing reduced the number of motorists even further and the few car owners I knew garaged their vehicles for the duration.

Buses fuelled by producer gas also appeared on our roads during the war. The gas was manufactured in a small two-wheeled trailer towed behind each bus by passing air over white-hot coke. The system used coke, produced in this country as a by-product of the gas industry, instead of precious petroleum that had to be imported in convoys at risk from U-boats. However, the calorific value of producer gas is very low and it seemed to be an ineffective system. Walking speed was all that they managed on Sticklepath hill.

Threshing was a contract operation, the machine travelling from farm to farm. The corn ricks were threshed one at a time when the grain was required for sale or to grind and feed to livestock.

Often the threshing machines arrived the night before and set up so that an early start could be made next morning. The hut and water tank were detached and the threshing machine drawn into position close beside the rick to be threshed. At our farm the entrance to the rickyard was on a sharp corner so that the engine and long threshing drum towed behind it had to make almost a U-turn to enter the narrow gateway. To enable the thresher to be precisely positioned, the engine had a hitching point at the front as well as at the back so that it could push as well as pull. To assist in manoeuvring, the rear axle of the thresher could be unpinned and steered manually with a pole that was inserted into a slot in the axle. Once in position close to the rick to be threshed the threshing machine was levelled with jacks and blocked up and secured with wooden blocks and wedges.

Once this was done the engine was moved into the

farmyard for the night, where the water tank for the boiler was refilled with a hose, and then it stood there, hissing softly for hours as it cooled. The thresher men went off home on the cycles they carried in their van although, I suspect that when they were too far from home they bedded down in the van for the night.

Next morning the engine was moved back into the rickyard and manoeuvred until its flywheel was in line with the driving pulley on the thresher. This ensured that when the endless leather belt driving the thresher was slipped onto the flywheel it would run true and not tend to ride off the flywheel or the pulley on the thresher. Once the engine was aligned, the belt was tightened by edging the engine back and jamming wooden wedges against the big driving wheels to prevent it moving. Meanwhile, the thatch had been stripped off the rick in readiness and everyone's duty allocated.

Two men came with the machines. One looked after the engine and the grain outlets from the thresher. He clipped a four-bushel sack under each of the four outlet spouts from the thresher. Two were for good grain and were filled alternately; one was for seconds and the fourth for 'tailings' and weed seed. When each sack of good grain became full he switched over to the other spout. Then he removed the full sack, tied the neck with a length of binder cord and clipped a fresh sack in its place. Usually, there were few sacks of seconds and only the odd one of tailings and weed seeds. The large four-bushel sacks were all hired from the West of England Sack Company and used solely for the transport and storage of grain.

The driver's mate stood on the platform on top of the thresher where his job was to spread out the sheaves evenly and feed the stalks head-first into the threshing drum through a narrow opening. This opening was the same width as the drum that did the actual threshing and removed the grain from the straw. Feeding the corn into the drum was a very responsible job. If the corn was fed in as wads, not all of the grain would be threshed from the ears and some would be lost over the back of the machine with the straw. Normally, the principal noise from the thresher was a steady continuous hum that could be heard at some distance, but whenever a wedge of material slipped in there was a momentary lower-pitched 'thrum' as it passed through.

The other key workers in the team were the pitchers, usually two, who pitched the sheaves from the rick onto the top platform of the thresher to the feeder's assistant. The pitchers together with the feeder determined the work rate for the whole team. The pitchers were usually young men, and when I worked full-time after leaving school I pitched at one or two farms. We started by climbing right to the top of the rick onto the ridge and started to pitch to the thresher, each being responsible for half the rick. At first it was easy because we were well above the level of the thresher platform, but after lunch, as we reached the bottom of the rick we had to stretch to get the sheaves up onto the platform. Sometimes, it became a contest between the pitchers when nearing the bottom of the rick to see who could finish his side first. Pitching was much easier if you understood how the rick was constructed and could scoop up the sheaves in the order that they had been laid in. However, if we pushed them up too fast and buried the feeder, sheaves would come tumbling back down, accidentally or deliberately, and we would have to pitch them up again.

The feeder usually had an assistant who cut the binder twine tied around the sheaf with a razor-sharp knife and pulled out the twine to prevent it going into the drum before pushing the loose sheaf to the feeder.

The rest of the team on the ground included two or three to take away the 'trusses' of straw and stack them, one or two men to cart away the sacks of corn, and one or two others to deal with the *dowse* falling out from the bottom of the machine. Dowse consisted of the chaff – tiny pieces of straw and dust.

The threshed straw coming over the straw walkers at the back of the threshing machine was bundled up automatically and tied with a band of binder twine at each end before being ejected from the machine. The twin knotters and ejection mechanism were identical to the ones used to tie sheaves on the binder.

These trusses of straw, called *wads*, were unlike the rectangular or cylindrical bales we see in the fields today. They were long and roughly oval in cross-section, and thicker at the middle than at the ends. They were easy to pitch because they were quite light but difficult to stack because of their shape.

Every few years, when reed was required for thatching, a field of a variety of wheat with long, straight straw suitable for thatch was grown. When this crop was threshed an additional reed-combing machine was included as part of the train towed behind the engine when it arrived. This was used to comb out all the bent and broken straws damaged during the harvesting operation or in the threshing drum, leaving the straight ones for thatch. The straw 'trusser' on the thresher that formed and tied the wads of straw was not used when reed combing. Instead the threshed straw was fed straight into the comber positioned immediately behind the thresher. The straight straws to be used as reed emerged from the machine in an endless parallel stream.

Several additional workers were required when reed-combing to tie up the reed into bundles we called 'niches'. Each worker took turns to collect armfuls of reed and lay them down carefully on a withy tie band that he had previously laid on the ground. Care was taken to ensure that all the straws were parallel and the butts level, although the ends of a central core of reed straw were allowed to project. When a niche was large enough it was tied as tightly as possible with the withy band. This was a thin length of willow with one end turned back on itself and twisted round the stem to form an eye. The other end was passed through the eye, bent round into a loop and twisted under the band to secure it. When the niche had been tied it was banged down onto a wooden platform laid out on the ground behind the machine. This drove the projecting wedge of reed into the base of the niche to tighten it and to give the niche a flat bottom on which it would stand. An average niche was almost 3 ft (1 m) in diameter and stood about waist high on the wooden platform. The bent and broken straw discarded by the reed comber was forked away and stacked loose.

Removing the sacks of grain into storage required one or two men depending on what was to be done with the grain. Four-bushel sacks of grain are heavy and difficult to handle. The bushel is a volume measure and so the weight of a four-bushel sack varied in accordance with the size and type of grain. Wheat is the heaviest and a sack weighed 2¼ cwt (approximately 114 kg) compared with 2 cwt (100 kg) for a sack of barley. Oat varieties varied considerably in density but on average a sack weighed about 1½ cwt (76 kg). Full sacks were taken to the barn on sack trucks. These had to be balanced with care as the sacks were waist tall and if the truck was leaned back too far the weight of the sack made it a struggle to right it.

At our farm the sack truck had to be pushed down the rough track to the gate of the rickyard at the right-hand bend in the road and then straight across to the barn on the other side. If the grain was to be sold, the sacks were set down upright, side by side on the floor of the barn. However, if the grain was to be kept and used on the farm for feeding the livestock, the sacks were emptied into hutches at one end of the barn. These were about 8 ft (2.5 m) square and were boarded over to form a loft above. The hutches were open-fronted but boards could be slotted into grooves at the sides to build up the front as the depth of grain increased. The first few sacks were trucked in with no front boards in position and emptied straight onto the floor. This sounds easy but, although some grain flowed out when the sacks were laid down and untied, the 2 cwt (100 kg) bags still had to be manhandled to induce the rest of the grain to follow. However, once the level of grain in the hutches had risen and several front boards had been inserted the rest of the bags had to be carried by a worker on his back up a flight of wooden steps to the loft above. Once up there the sacks were emptied through hatches into the hutches below.

Although workers took pride in being able to carry these massive weights up the stairs, it was a dangerous business and many ruptures and back injuries resulted, often not becoming apparent until years later. The knack was to carry the sacks very high on the back and bend forward from the hips to achieve balance. Two men were required to raise the sack to a position high enough for the carrier to take it up onto his back. The coarse sacks were too thick to dig one's fingers into to get a grip at the bottom. So each man held the top loose corner of the sack with one hand, and one end of a wooden sack bar positioned near the bottom of the sack with the other. The sack was leaned forward so that the weight could be shared equally by each hand, and then it was a straight lift high enough to edge the bottom onto a suitable waist-high surface, followed by another struggle to rear it to an upright position. Of course, each man was still lifting over 1 cwt (50 kg) every time a sack was lifted. The same procedure was used when grain to be sold was loaded onto a lorry. I lifted many such sacks in this manner but was never required to carry them upstairs.

The last members of the threshing team worked almost under the machine pulling back the dowse. This was either bagged or dragged back and heaped to be dealt with later. Ultimately, it was given to the poultry for bedding and to allow them to scratch around and peck out any seeds still in it. Removing the dowse was probably the least pleasant task associated with threshing. The dust was thickest there right next to the bottom of the throbbing machine and it was noisy. The sieves that separated the grain into grades were supported on external hangers that shuttled to and fro just above your head and several drive belts ran nearby. There was no guarding, so you had to look out for yourself and be careful. However, as it was physically undemanding, it was my first job soon after I started to spend time on the farm, usually in the care of one of the women helping out for the day who ensured that I kept clear of the moving parts. The only other jobs occasionally done by women on the farms where I threshed were acting as the feeder's assistant – cutting the bands on top of the thresher and helping on top of the straw rick.

Threshing required a team of nine or ten workers in addition to the two that came with the thresher, and yet more were needed for reed combing. This was more than was available on the average farm and so workers came from neighbouring farms to help. In return, when they threshed, someone would be sent to help them. At my second farm I was the obvious person to send and I went to several nearby holdings. I went on my bike with my pick strapped to the crossbar, the tines projecting dangerously forwards in front of me.

Although threshing was very hard work, carried out in very dusty and sometimes very hot conditions, it was considered to be something of a day out. Instead of going home for lunch or taking a packed lunch, the whole team went into the farmhouse for the midday meal. At every farm that I visited the kitchen was large enough for the whole team to sit down together around a long, refectory-style table. We sat on benches rather than chairs, and were waited on by the farmer's wife who had also prepared the meal aided by her daughters or friends.

I can only remember one menu on these occasions, a generous slice of meat pie with masses of vegetables, followed by apple pie and clotted cream. The morning's work had always given us large appetites and we all consumed the generous portions we were given. Even after the end of the war, when my employer loaned me out to other farms, rationing was still in force but most of the food was home produced and there was always plenty. It was always a friendly occasion providing an opportunity to catch up on the local news and gossip, and there was much chatter and banter, particularly when the farmer had grown-up daughters helping their mother. However, there was no dallying after lunch and we went straight back to work.

The last time that I threshed, the traction engine had been replaced by a tractor. There were few tractors in the country powerful enough and heavy enough to pull a threshing machine and to drive it, so an American MM was used, but Field Marshalls were also beginning to be used for this purpose.

Once the combine superseded the binder the thresher almost disappeared and was only occasionally brought back into use when it was needed to thresh and comb a crop of long-strawed wheat grown to provide reed for thatching, an operation that a combine cannot do.

The grain that was retained for feeding to livestock was ground with a grist mill. This stood in one of the barns and was driven by a belt from the tractor that stood outside. Grain to be ground was poured into a hopper above the machine and flowed by gravity down into the grinding mechanism. This consisted of two vertical circular plates, one stationary and the other rotated by the drive from the tractor. The plates had grooves incised in their surfaces that ground the corn as it passed between them. The fineness of grinding was adjusted by a screw that varied the tension on springs pressing the plates together.

Chapter 9

Livestock

NORTH Devon was, as it still is, predominantly a livestock area. The mild, wet climate encourages the lush growth of grass but the high rainfall coupled with the undulating terrain with steep slopes and small fields does not lend itself to arable cropping. This becomes more marked towards the high land of Exmoor and further to the south approaching Dartmoor. Local breeds, such as the North Devon 'Ruby' cattle with their long, straight or slightly forward-curving horns, the Devon Closewool and Exmoor Horned sheep were developed to suit the conditions. These breeds are largely confined to North Devon, although the relatively small but compact reddish-brown cattle have been introduced to other areas of the country because of their ability to gain weight rapidly. The Devon Ruby was prized as a draught ox on early medieval farms when oxen were commonly used as a favoured alternative to horses for pulling ploughs and hauling wagons. Oxen had the advantage that in addition to their role as draught animals they provided milk during their lifetime and a roast at the end of their days.

During the war farmers were required to plough up

grassland in order to plant food crops and as a result both farms on which I worked had larger areas of arable land than before 1939. This temporary shift in emphasis had a major effect on my choice of agriculture as a career. I soon realised that my chief interest lay in field work, particularly mechanisation.

After the war most farms in the area reverted to stock rearing and milk production for which they were ideally suited. There was a brief period in the 1980s when the price of cereals led to large tracts of grassland being ploughed up again in order to grow cereals. Unfortunately, this included areas of marginal land and high land with thin soils and high rainfall totally unsuitable for cereal production. Much of this had not been ploughed in living memory, if ever, and cultivating it was only made possible by the use of modern high-performance machinery. This change proved to be short-term and most of the land has now reverted to livestock farming.

As both my farms were in the valley of the river Taw they were in the part of North Devon that could support some arable farming, and both would have been described as 'mixed' farms even before the war. Both had cattle and flocks of sheep as well as growing a range of crops.

Both farms had dairy herds that were of average size for the time, although they would be considered small in comparison with the specialised dairy herds of today. When I started, the herd at my home farm included a mix of breeds. There were several Shorthorns – brown cows with brindled patches where white hairs were mingled with the brown. They were considered to be a dual-purpose breed, that is reasonable milk producers that also fattened well. Shorthorns are rare now and dual-purpose breeds have largely gone out of favour in deference to single-purpose breeds that are either good milk producers or fatten well. There were also one or two of the North Devons that are fattening cattle but not valued as milk producers, although their milk is rich; several Ayrshires that are the opposite, and a few attractive yellow-gold Guernseys. The last were largely responsible for the rich head of cream on the milk that customers on the milk round appreciated and desired. The herd was gradually changed to be all ginger-brown and white Ayrshires with their distinctive up-turned horns because of their high yields. In turn, these were replaced with the even higher-yielding black and white Friesians. My employer at my full-time workplace had a pedigree herd of Friesians when I joined him.

On both farms the cows were milked in cowsheds we called 'shippons'. These were of a standard pattern that was repeated in contemporary cowsheds all over the country. The cows stood side by side in a row along the length of the building and were separated into twos by waist-high concrete partitions far enough apart for there to be room for us to walk in between each pair to milk them. Each animal was loosely chained by the neck to a steel bar bolted to the partition next to it. The chain was free to slide up and down on this bar so that the cow could stand or lie down as she wished. Each cow had a water bowl she could operate with her nose by pressing down a hinged plate in the bowl that opened a valve allowing water to flow into it. There was a common manger across the front of each stall for their food with a walkway, the 'feeding passage', which ran the full length of the shippon beyond the manger. We carried the fodder to the cows along this passage. Behind the row of cattle there was a shallow dung channel that ran the full length of the building to collect the muck and urine, and beyond that a wide passage in which we worked and which the cows used when they entered and left the building.

At every milking throughout the year each cow in milk was given 'concentrates' in the form of cow cake that was put into the manger in front of her. Oil seed cake was bought in from cattle feed merchants, and was formulated for different times in the year to provide a balanced ration for milk production in conjunction with the hay and roots fed in winter and with the grass grazed in summer. The cake was supplied in the form of 1 in. (25 mm) cubes and was rationed out to each cow in proportion to her yield. In summer the grass provided a greater part of the cow's nutritional requirements and so less cake was fed. When a cow's yield had diminished to almost nothing she would be dried off and the cake allocation stopped. The cows loved their cake. It was put into each manger just before the milking machine was put on to occupy them and, much more important, to stimulate the let-down of their milk. Other regular routines, such as washing the udder, also had the same effect in helping to ensure rapid and complete release of the milk.

The cows were milked by machine using three individual bucket units. These were operated by a vacuum produced by a vacuum pump sited next door to the shippon. A vacuum pipeline ran along behind the stalls at a high level with taps spaced behind each pair of cows. The milking buckets were connected to these taps by rubber tubes so that three cows in different stalls could be milked at the same time.

We started by washing the cow's teats and udder with a cloth dipped into a bucket of water with disinfectant added. Then, just before a cow was attached to a machine she was given her cake ration and a single squirt was milked by hand from each teat into a 'strip' cup to check that no mastitis was present. If clots were visible on the black plastic top of the cup the milk from that teat was discarded. If all clear the four teat cups were slipped on. The teat cups were metal tubes each containing a pre-formed rubber liner. When I started only plain tubular liners were available and they had to be shaped before being fitted into the teat cups by inserting a metal ring into the tube with a special tool.

To attach the machine the cluster from which the four teat cups sprouted was held in one hand with the four cups dangling down. In this position the vacuum was closed off so that the cups could not suck in dirt. Then each cup was fitted onto the appropriate teat with the other hand – first the front two, followed by the back two. When each cup was up-turned the vacuum was restored and the suction held the teat cup on the teats. A 'pulsator' automatically varied the vacuum to the space between the rubber liner and the outer metal wall of each teat cup, alternately squeezing and releasing the teat to produce the milking action and induce milk flow. The action closely resembled that of hand milking. The milk flowed under vacuum through a rubber pipe from the teat cups to the sealed bucket.

When the cow had released all her milk the vacuum was turned off and the teat cups removed. Once the vacuum ceased they almost fell off. Then the teat cups were transferred to the other cow in the stall and when she had finished the milking machine was disconnected from the pipeline and moved on to another stall.

When the bucket was nearly full the lid with the pulsator and all the pipes attached was transferred to a spare bucket and the bucket containing the milk was taken to the dairy next door and emptied into a reservoir above a milk cooler.

The cooler was a narrow tank through which tap water flowed. The milk was cooled by allowing it to flow from the reservoir at the top down over each side of the cooler, the sides being corrugated to increase their surface area. The cooled milk passed through a filter into the churn in which it was to be stored before being taken on the milk round. The degree of cooling achieved by this method was limited by the temperature of the tap water available.

Refrigeration was not available for a unit of our size and so the evening milk was stored overnight in the dairy at room temperature before being taken onto the milk round next morning. The milk from the morning's milking was treated in the same way except that it went straight onto the round. When I started, none of our customers had refrigerators and so they kept the milk in the larder on a slate shelf or outside in a meat safe – a small box with perforated zinc sides that allowed air to circulate. It is not surprising that from time to time they complained that their milk had gone sour. Naturally, this was most common during summer when night temperatures were high and because there was little protection from the sun for the churns on the trap, or for bottles in the van that we eventually used.

Our milk was not pasteurised; that was very much a factory process as it usually still is, and our cows were not attested for freedom from TB or from brucellosis. These procedures that are now a requirement were still in the future. Today we seldom stop to appreciate that our milk comes from herds free of these diseases and is refrigerated from the moment it leaves the cow until it reaches the supermarket, having been pasteurised in a factory en route. Even at the supermarket the racks are chilled with cold air circulated around the crates. These advances make the extended 'use by' dates printed on the containers possible. Our milk would have had to have been accompanied by a label stating 'use today'.

Some hand milking was done at my home farm but it was confined to freshly calved cows that still had their calves with them and were kept apart from the main herd. Hand milking required very strong wrists. The knack was to squeeze the top of the teat firmly with the thumb and first finger to trap the milk that had let down into the teat cistern, and then to squeeze the whole teat with the other fingers to expel the milk into the pail. Both hands were used to milk the front two teats alternately, and when these had been drained the back ones were milked in the same way. George was a very skilled milker. Before the milking machine was introduced he had had to milk his quota of cows before going to school. If one of the farm cats appeared when he was milking, his party piece was to squirt a stream of milk at the cat and his aim was so good that all the cat had to do to have its tea was open its mouth.

The hand milker sat on a low three-legged stool just beside the cow, facing backwards with the head pressed against the cow's flank. Before sitting down one reversed one's cap so that the peak was at the back. Then it was advisable to mutter soothing appreciative things to the cow to put her at her ease because, as I soon found out when I started hand milking, the milker is in an extremely vulnerable position if the cow decides to kick. A cow normally kicks forwards and can easily unseat the milker, upturn the bucket and raise bruises or cause pain in most embarrassing places. However, kicking was not a frequent occurrence. Usually, it was because a teat was sore or because the animal was a heifer that had just given birth to her first calf and the whole business of being milked was new to her. After a day or so most heifers settled down. Nonetheless, there were examples of animals that possessed a nasty streak or possibly recollected some hurt suffered in the past and would regularly kick out when they were being cleaned, and even try to kick off the teat cups of the milking machine. Washing and fitting the teat cups could be done in relative safety by leaning against the back leg facing forward so that if she did kick, it was only the wrist and arm that were in the way. In some cases where it was a regular occurrence we looped a cord around the cow's hocks, crossing it between the legs to hobble her and prevent her from kicking. Habitual vicious kickers were sold.

The cows stayed in the shippon all day during winter. At first in late autumn they were kept in

Mr Avery cutting corn with a horse-drawn binder in 1930 near Georgeham. Mr Philip Avery is leading the horses. The horse in the centre is smaller than the two flankers. Just the year before I started at my first farm a similar team of horses was used in the binder with Punch the milk-round pony in the middle. Braunton Museum.

Cutting corn with a binder. The tractor is a Fordson on steel wheels with spade lugs on the rear ones. Road bands would have had to be fitted before it could leave the field. The sheaves standing against the hedge were probably cut by scythe and tied by hand to 'open up' the field for the binder. Beaford collection.

Cutting corn with a converted horse binder like ours. Beaford collection.

Pitching sheaves two at a time onto a tractor-drawn trailer in a local cornfield. The lades are identical to those described for horse carts. North Devon Athenaeum.

Wheat sheaves stooked in eights near Riddlecombe looking towards Exmoor across the valley of the river Taw (above). As soon as combines replaced binders straw was baled and the stubble fields looked quite different (below). Ravilious collection.

The rickyard at Buckland farm. It co
have been ours or almost any other in North Devon althoug
we climbed up onto the cart to pitch off th
load and never used
ladder. Note the leng
of the rick ladder tha
we also used for app
picking, and the propping of the completed rick that s
has to be thatched.

Museum of Barnstaple and North Devon.

A Claas trailed combine pulled by a Field Marshall in a field of barley at Halmpstone farm near Bishops Tawton in 1951. North Devon Athenaeum.

A bagging combine working at Ashwell. The second operator on the machine at the rear is bagging the grain and dumping the bags, my first role when we began to combine. However, judging from the licence plate, it is working a decade after I left my farms and is a more recent model than the Massey Harris 726 we hired. Ravilious collection.

Baling on the home farm with the New Major and International B45 baler. Author's photograph.

Loading mangolds at Goodleigh in 1953. The unusual butt has rubber-tyred wheels instead the normal large wooden wheels, the small wheels allowing a wider body overhanging them be fitted. The horse's harness described in the text is clearly shown. North Devon Athenaeum.

Barley threshing at Barton farm, near Braunton in September 1949 with a Field Marshall drivi the thresher. The partly threshed stack and the presence of the woman with the basket suggest that it is a drinks break. The team is larger than the ones with which I worke Note the folded West of England grain sacks piled on the ground. North Devon Athenaeum.

Threshing with a steam engine driving the thresher.
The engines that came to us were all roofed, unlike this one. Beaford collection.

Reed combing at Westacott, showing the adaptation to the thresher and the niches of reed on the wooden platform. The model of the tractor on the right indicates that this photo was taken some years after the others. Ravilious collection.

Drilling corn with a pair of horses. Beaford collection.

Cutting grass with a pair of horses near Barnstaple in 1942. Note the different harnessing arrangement when using draft chains instead of shafts.

Museum of Barnstaple and North Devon.

A David Brown Cropmaster pulling a tedder. North Devon Athenaeum.

Carting hay with a Standard Fordson providing the pulling power.

Museum of Barnstaple and North Devon.

Barnstaple cattle market. Museum of Barnstaple and North Devon.

A young farmers' shearing competition at Cobbaton in 1952. Note the hand-operated shears. North Devon Athenaeum.

The upper yard at the home farm in about 1955. The cart linhay is to the le and the farm house and fenced front garden at the far side. The thatch had been recently patched by laying a new layer of reed on top of the worn thatch The roof on the other wing of the hous behind also may have been repaired. A set of spring harrows is leaned against the wall and a scuffle stands at the edge of the linhay.

Author's photograph.

overnight and grazed in the fields during the day. Then, as winter advanced, when the grass growth slowed and the fields began to poach, they were kept in all the time. When spring returned and the grass began to grow again they were turned out once more, at first just during the day and then all the time, only being brought in to be milked. They loved being out for the first time and normally sedate, slow-moving cows would throw up their heels and jump about like young calves.

When they were lying in, there was a lot of extra work in looking after them. They were fed hay twice a day and chopped mangolds, and after each milking they were cleaned out. This involved pulling back any dirty straw into the dung channel and then clearing out the channel. The muck was shovelled into a wheelbarrow and wheeled outside. The channel was then washed down and brushed out, and straw was spread under the animals so that they could lie on it. The muck was stored in a heap in the 'midden' where it gently steamed and rotted before being carted out to the fields in the spring and spread as a useful source of plant nutrients and for its valuable soil-conditioning properties.

In the summer the cows were only brought in for milking, and as they were in the shippon for a short time there was little cleaning to be done. However, when the cows were first let out the grass was very lush and this made them very loose so there was a tendency for them to make a mess. When they did, it went everywhere, right across the work passage, onto the walls and onto us as well if we didn't move out of the way quickly enough.

A much more serious consequence of them gorging too much lush grass was that they might become 'blown', seriously bloated by gases formed in the stomach. This could be fatal, and the remedy was to puncture a hole into the stomach with a veterinary tool to release the gases. I never witnessed this apparently crude but effective remedy although I understand that in desperation a farmer might do this himself if he couldn't obtain a vet in time. To prevent this happening we only allowed the cows to graze for short periods on fresh grass at the beginning of the season.

Fetching the cows from the field for milking was an easy, pleasant task on a dry, warm summer's day. They were usually ready to be milked and would start to move towards the gate as soon as they saw you appear and start to call 'co-oh, co-oh'.

Although traffic was light on the minor road beside the farm, George's father also rented pasture on the far side of the Barnstaple to Bideford main road. When bringing them home from these more distant pastures, the cows had to walk a quarter of a mile along the main road before they turned down a lane onto our land.

Even though the volume of traffic was much less than today we caused chaos on Saturday afternoons in the summer, as all the holiday traffic leaving North Cornwall travelled along this stretch of road, and long queues of vehicles formed up behind us. Such queues were not the everyday occurrence that they are today, and drivers often became irritable and would try to drive through the herd if we didn't prevent it by bunching the cows and walking in front of offending cars. We discouraged cars pushing through the herd as of course there were also cars coming towards us, the road was narrow and it was easy for cows to be wedged in between cars and be injured. Not only was this dangerous for the cows but it was hard on wing mirrors and the bodywork of cars. However, the cows seemed to enter into the spirit of the adventure and regarded an open window as an invitation to peer in, usually much to the consternation of the occupants. We did come in for some verbal abuse, but there was no alternative way of taking the herd back to the farm and the majority of motorists were patient.

When I started to work full-time on my second farm after leaving school I became responsible for milking the herd of cows there. I milked morning and afternoon, Monday to Friday and on Saturday morning. My employer usually did the rest of the weekend milking. The morning milking was started as soon as I arrived at 7.30 am, and in the afternoon I returned to the farm from the fields to start again at 3.30 pm. There was no milk round, and so the milk was collected each morning by lorry. The milking had to be completed, the milk cooled and the ten-gallon churns lugged out and heaved onto the milk stand by 9 am each day. Ten gallons is equivalent to about 45½ litres, and a full churn weighed almost 50 kg. Hoisting a churn onto the stand when I was on my own was a struggle as the handles at the top just under the lid were at waist height before starting to lift. Moving them was not so difficult as I soon acquired the knack of tilting a churn and balancing it on the bottom edge. Then, using the handles, the churn was rotated so that it rolled along, rather like trundling a child's hoop.

The milking routines were identical with those that I had learned at my home farm under the tutelage of George, the only difference was that there was no mains electricity, and so the milking machine pump was driven by a stationary engine next door to the shippon.

The engine had quite high compression and it was almost impossible to turn it over with the starting handle. To start it a decompression valve had to be held down with one hand while the engine was cranked with the other. Once the engine had been cranked up to a reasonable speed the valve was released and with any luck the engine would start. Fortunately, the engine was very reliable and it never failed me, although I always had my heart in my mouth when I went to start it, particularly when my employer was absent from the farm because there was no alternative means of milking the cows. Cows that are used to being machine milked do not take kindly to being hand milked, and the Friesians had small teats that were difficult to hand milk.

I quite enjoyed doing the milking each day, particularly in the wintertime when the weather outside was bad. Having cycled for half an hour up and down hills to get to the farm in the wet and cold it was a welcome relief to go into a shippon that was lovely and warm from the heat given off by the cows.

During the winter that I worked full-time it seemed to rain first thing almost every day, although it often cleared by mid-morning. One of my mother's old saws was, 'rain before seven, shine before eleven', and I usually found this to be a reliable forecast. My grandfather had a trawler in Dorset and Mum had gained her weather lore from him. I wish I could recall more of her sayings. Another that most country folk know and is equally reliable is, 'red sky in the morning, shepherd's (or sailor's) warning, red sky at night shepherd's (sailor's) delight', although mother recognised an 'angry' red sky at night that belied the prophecy and was not the precursor of a good day. Equally well known and dependable was, 'long foretold, long last; short notice, soon past'. Yet another trustworthy favourite of hers that I never remember anyone else quoting was 'a southerly glen (or glim, I cannot remember which) and a wet shirt', meaning that a brightness along the southern horizon during or just after rain indicates that there is more to come, and soon.

We had no milk delivery round at this farm, and so as soon as the milk churns had been trundled out ready for collection my next task was to wash up all the milking utensils. This was not the most pleasant job as the detergent used on farms at that time contained sodium hypochlorate and this was hard on the hands. I found that open cuts and wounds on my hands took ages to heal and I always seemed to have open, raw sores. No one used gloves in that era and I recall that George suffered with dermatitis that he blamed on the cleaning fluids.

There was a bull at each farm that was kept in order to bring the cows and heifers into calf. At my first farm the bull was housed by himself in a small lean-to building opening onto the inner yard. When there was a cow to be served he was released into the yard with the cow and left there to do his duty. At first, when a teenager, I was not allowed near the yard when the bull was busy there. Naively, I thought that this was solely because the bull was dangerous, and it was some years before I saw what went on and realised that the farm staff had been shielding me from witnessing sex 'down on the farm'.

At the farm where I worked after leaving school we relied on artificial insemination. This was intended to improve the quality of our pedigree herd by providing us with the services of a high-quality sire. To retain a bull would have been much more expensive than my employer could afford or justify with the small number of animals in the herd. Unfortunately, AI was not very successful and many of the cows did not come into calf, resulting in too many dry cows in the herd and a sharp drop in milk production. In desperation my employer purchased a Devon bull to see if this would improve the fertilisation rate, which it did.

As at my home farm the bull and cow in season were released into the inner yard. One side of the yard was enclosed by the boundary wall of a small cottage attached to my employer's farmhouse. It was occupied by an elderly man and his wife. The old man used to object to what took place within the yard and from behind his wall would wave his stick at us and shout 't'idn't daicent' in the quavering voice that some ancients have. I am ashamed to say that both my employer and I were greatly amused by this regular occurrence. We ignored it as there was nowhere else that the bull could be put safely with the cow.

I did have one scary encounter with this bull. During the summer he was kept in a field with a group of young heifers in order to get them into calf for the first time. One Saturday morning, not long before I left the farm, I was clearing up ready to go home when the phone rang. I was on my own as my employer and his wife were away and wouldn't be back much before the afternoon milking. When I answered the phone I was told that our bull had broken out of the field and was making his way on his own in the direction of the village that was about a mile distant. I had no alternative but to go and fetch him and shut him up in his house as he might have broken out again if I had returned him to the field.

I went on my cycle to find him, and then drove him home at a smart trot as I was keen to finish work and go home. When I got back to the farm I realised that this had been rather foolish and short-sighted. I was faced with a very hot bull who was obviously not in the best of humour, and I had to put him into his house, and then go in with him and chain him up. To do this I had to push him up against the wall, stretch over his neck to unhook the tether chain from the wall, and then reach under his neck to grab the other end so that I could link the ends together round his neck. My heart was in my mouth throughout, and I breathed a sigh of relief when he was safely tethered. According to my family records, a bull killed my great-great-grandfather when he was a great age, 100 according to one version, mid-nineties in a second. I had no wish for this to become a family tradition, at least not until I was older than he was. Fortunately, our bull was a small North Devon and beef breeds are usually much quieter and more reliable than dairy breeds. I doubt if I would have got away with it if he had been a big Friesian, or even a Jersey as that breed has a bad reputation despite its relatively small size.

There was a large flock of Devon Closewool sheep on my first farm. The Closewool only became recognised as a separate breed in the 1920s, having originated from a cross between the Devon Longwool and Exmoor Horned breeds, and it is restricted in the main to the North Devon area. It is a stocky square breed of sheep with curly wool so dense and tight that it is hard to dig your fingers into the animals to get a grip to hold them.

They were quiet and easy to manage compared with some of the hill breeds that I have encountered since. To catch one they were bunched into the corner of the field and then it was easy to push in amongst them to grab the one that needed to be inspected. Having caught the sheep, it could be held with its neck between your legs so that its back and sides could be examined. Alternatively, the sheep could be turned and sat on its bottom with its legs facing away from you so that its feet and belly areas could be examined. The first position was easiest to achieve, but it was possible for a lusty animal to push forward so hard that you went with it riding on its shoulders, a situation that was hard to retrieve without losing one's dignity.

We had two sheepdogs at my first farm. One was a Scottish Collie who did most of the work. The other, an Old English Sheepdog, was lazy and usually managed to disappear when work was to be done. It was difficult to persuade him to accompany you to the field. Having got him there it was even harder to persuade him to collect the sheep and you did most of it yourself. If you turned your back on him to deal with some problem affecting an individual sheep, when you looked around he was a distant shape disappearing through the gate on his way home.

Unfortunately, the dog had been given the same name as myself. This gave endless pleasure to one of the older farm workers who used to call out whenever both the dog and I were nearby, 'Michael, you filthy dirty thing, you, cum yer,' or something similar, often less politely worded.

This man was a strange character who had been born in one of the farm cottages and still lived in one. He was a very skilled farm worker who always seemed to be given all the pleasant jobs; often they were long-established perks. For instance, when corn harvesting, he spent the time sharpening the knives and never did any stooking, although he was the worker who scythed around the field before the binder could start. He had been the principal horseman but he never transferred to the tractor; he probably would have liked to have done, but he had no driving licence. Furthermore, the tractor was George's responsibility, although he wasn't particularly keen on driving it for long periods and so he delegated it to me whenever he could. As I got older and became more experienced I did more and more of the tractor driving during school holidays and at weekends. This continued when I was home from university and as the long vacations coincided with the busiest field work periods of the year I still did the bulk of the tractor work.

I recall vividly that this led to an altercation with this particular worker in a silage field about who should drive the newly acquired Fordson Major that I had assumed, correctly as it turned out, would be my job as it had been with the old Standard Fordson. By this time George was the farmer, and I think that the worker resented the younger man, whom he had known from a child, being the boss and taking decisions that he might have taken in the time of George's father, when he often acted as foreman. He became more and more difficult and awkward and fomented trouble with the other workers. Ultimately, this came to a head, and I can still remember walking into the farm kitchen one morning to be told with great glee by George's wife that the man had given in his notice.

George did much of the shepherding. He was particularly good at helping the ewes when they had difficulty in lambing. Occasionally, a lamb was presented the wrong way round or with its leg or head bent back, or if there were twins with their limbs intertwined and coming forward together. He could gently move a lamb back and turn it into the correct position for the ewe to give birth naturally.

The other major task associated with the sheep was shearing. This was done mechanically with clippers rather like those used in a barber's shop. The hand-piece consisted of a blade that reciprocated across a similar fixed blade, in principle rather similar to a binder cutterbar in miniature. Both blades were equipped with narrow, pointed fingers that did the actual cutting with an action rather like a whole series of scissors working side by side. The blade that reciprocated was driven by an articulated drive shaft with two universal joints that allowed the shearer to move the hand-piece freely in almost any direction.

At my first farm the power was provided by an

electric motor suspended from a beam. At the second farm there was no mains electricity and the power was provided manually by cranking a handle on a gearbox mounted waist high on a steel upright with a tripod base. Turning the handle was my role, and I was expected to maintain a steady continuous crank speed. It wasn't hard to do this as the handle wasn't particularly hard to turn, although the job did become tedious after a while. However, it provided me with the opportunity to watch and learn how to hold the sheep in a position that was comfortable for the sheep so that it didn't struggle and allowed the shearer to remove the fleece as quickly as possible. The shearer took the first cut down the full length of the back and then removed each half of the fleece in turn by making long, sweeping cutting strokes from the back right around to the belly.

Although I never actually sheared, my role was to catch the next animal to be sheared and to roll up the fleeces. The two half-fleeces were laid out flat on the floor, one on top of the other, the ragged edges and any bits were folded in and then the fleece was rolled up into a ball and tied. The rolled-up fleeces were then packed tightly into huge square hessian bags which when fully packed were sewn up ready for collection.

At the first farm shearing was done in one of the big barns, the centre of which was equipped with a wooden shearing floor. The sheep were brought in from a pen just outside the door. Invariably, someone joked that Michael, the scruffy longhaired sheepdog, would benefit from being sheared. Somehow he always seemed to sense this, and he was never to be found when shearing was in progress.

At my second farm there was no shearing floor in a barn and the small flock was shorn outside in the orchard, on the grass.

During the rest of the year, the sheep were checked and counted daily, more frequently at lambing time. The problems we looked for were lameness and, in summer, the presence of maggots. Lameness was usually caused by a stone or twig trapped between the claws or claws that had grown too long. The last was simply treated by paring away the horny growth, a painless operation no different to cutting one's own nails. Fortunately, neither flock suffered with serious foot rot, possibly because the fields grazed by the sheep either sloped or were well drained. Foot rot can be a serious problem on wet grassland.

Maggots were unpleasant. They resulted from eggs laid by blowfly in the dense wool of the sheep, particularly the dirty wool near the tail. The maggots that hatched ate into the flesh and caused raw open wounds. In the summer when this was a routine problem we kept a watch for animals that were scratching at themselves with a rear leg or twitching their wool or, in really bad cases, looking sick and ill. When a sheep with maggots had been spotted and caught, we rubbed the maggots out of the wound, trimmed away the wool around it with hand clippers and treated it with a disinfectant. This was a grey opaque liquid with a strong unforgettable smell that we carried in an old lemonade bottle. I never found out what it contained but I suspect that it was mixed on the farm from some recipe probably long since forgotten. Whatever it was, it was effective and the wound soon healed, although the sheep was checked every day until the wound had healed completely.

We also trimmed any dirty wool with hand clippers, an activity we called 'dagging', in order to discourage the blowflies from laying their eggs. The clippers had two triangular blades fixed to a U-shaped spring handle so that the blades crossed with a scissors action every time the handle was squeezed. Almost identical shears are still available at garden centres for trimming grass, although the blades are ground to a shallower angle for shearing wool – 45 degrees compared with 65 degrees for grass. Using these for any length of time requires very strong wrists just as hand milking does, and yet only a few years before I started there were no mechanical clippers and the whole flock was sheared by hand.

I gained little experience of pigs. At my first farm two or three were kept in an old-fashioned sty in a small walled enclosure, but the farmer's wife looked after them. They were destined to be slaughtered and the meat cured for home consumption. I can recall the hams suspended from hooks in the ceiling of the walk-in larder. There were no pigs at my second farm.

The farmer's wife also looked after the farmyard hens. They made most of their living by scratching around the farmyard, even venturing as far as the rickyard where the pickings were better, and laying their eggs in odd corners and in the bottoms of the hedges.

She also kept ducks and geese that she reared for the Christmas market. A week or so before Christmas the long kitchen was turned into a workroom where the farmer's wife, George's wife and one or two helpers killed, plucked, cleaned and trussed innumerable birds. Anyone venturing in there was likely to have a bird thrust into their hands to pluck. One soon learned to avoid the kitchen during this hectic time, but before doing so I did acquire another useful skill.

Both farmers ventured into poultry, keeping the birds in deep litter houses and in battery cages while I was with them. The cages were larger than is often the case today and only one bird was kept in each cage. My only lasting memories of those early cages were the number of eggs that got broken by the birds and the mechanical problems associated with the hand-cranked gear for cleaning the floors of the cages. Scraper bars attached at each end to endless chains were pulled along the length of the battery when the handle was turned. The stress on the chains was considerable and they often snapped. The birds in the deep litter house had plenty of space to move around and scratch although they did not go outside.

Chapter 10

Spring

SPRING was my favourite season in the Devon of 50 years ago. The roadside and field banks were studded with flowers and the air was full of birdsong. I wish that I could list the names of the flowers that sprinkled the roadside banks and verges. But, unlike most other country-bred men and women I have a limited knowledge of wild plants and can recognise only a few of the more common ones. I have always had difficulty in identifying and remembering the distinguishing features of plants and my wife, a keen and knowledgeable plantswoman, despairs of me when I fail to identify the garden plants that she cultivates. When at university this mental block proved to be a major problem as I was expected to learn to recognise common weeds of crops and to be able to identify cereals and grasses at different stages of growth. I managed to cope with this and to meet the course requirements but as a result I am now able to identify more grasses from their flowering heads than I can wild flowers.

Bird recognition was different. I was always keen to be able to identify the birds in the garden and in the fields next to my home, and one of the earliest gifts that

I can remember receiving was a bird book given to me by my godfather who lived next door. When we went for the regular after-lunch walk on a Sunday afternoon, my father and I were always on the lookout for bird's nests, and in those days we found many. As I got older and ventured forth on my own into the fields and lanes, I scoured the hedges and verges for nests. I became adept at spotting them and identifying the builders of the commoner ones from the construction of the nest and the colours, size and markings of the eggs. One of my clearest boyhood memories is of cycling with the school natural history society to Arlington Court, now a National Trust property, to see the heronry and to walk the woods. Then it was still in private hands and not normally open to the public as it is today, and so to be able to visit was a rare privilege. I can still recall hearing a woodpecker drumming, a much less common sound today, unfortunately.

Sadly, the wide and often indiscriminate use of herbicides and pesticides and changes in farming practices have wrought major changes in the flora and fauna of North Devon, although perhaps less so than in many other parts of the country. There is a visible reduction in the number of flowers and the number of flower species, and there are far fewer birds to be seen and consequently much less birdsong. It is still a wonderful place to be in the spring, but it has lost some of its charm and there seems little likelihood that the changes will ever be reversed.

As winter turned into spring little changed at first on the farm. Hedging and the routine care of animals continued as it had throughout the winter. Grass only stops growing for a brief period in the river valleys of North Devon and so one of the first jobs in early spring was to apply fertiliser to encourage vigorous and rapid regrowth.

This was done with a fertiliser distributor, a horse-drawn machine about 8 ft (2.6 m) wide. The machine consisted of a full-width, low-level hopper on wheels, with shafts for a horse. The axle between the wheels passed through the hopper and was equipped with projecting knobs so that it acted as a simple agitator, ensuring that the fertiliser flowed freely and uniformly through the narrow delivery slot that stretched the full length of the hopper. The application rate was changed by altering the width of the slot by means of a handle on the hopper. Fifty years on, most of the hand-pushed fertiliser spreaders sold at garden centres for the application of lawn fertiliser work on exactly the same principle.

To ensure that the grass grew uniformly over the whole field the fertiliser had to be applied evenly. This required the distributor to be driven so that the wheel just overlapped the previous wheel mark. We drove backwards and forwards across the field making a U-turn at each end. The horse between the shafts at the centre of the distributor could not be expected to walk straight up the field exactly 4 ft (1.2 m) from the previous wheel mark, particularly as the bridles were all fitted with blinkers that limited sideways vision. In order to achieve the correct degree of overlap, it was therefore necessary to guide the horse precisely with the ropes used as reins, one tied to each end of the horse's bit. This was quite difficult to do on grassland where the wheels scarcely left any indentation and the marks were barely visible even to the horseman walking behind the distributor. The same machine was also used to apply fertiliser on arable land, but in the soft cultivated soil the wheels dug in and left visible marks that were easy to follow and made it simpler to maintain the correct overlap.

When the hopper was almost empty it was re-filled from sacks carried to the field on a cart. This was the most arduous part of the whole operation as the industry still used 1 cwt (50 kg) sacks. While working full time at my second farm I was expected to load these onto the cart by myself. This was very hard work, especially when the last sacks lying flat on the floor of the barn had to be lifted up to waist height before I could load them onto the cart.

Most of the fertiliser that was applied was in granular form although these were not as uniform as prills, the spherical particles that are often supplied today. The fertiliser was delivered in either paper or hessian sacks and often became damp when stored for any length of time in the barn. The fertiliser had frequently become lumpy by the time it was used and these lumps had to be broken up when the hopper was refilled.

Every few years lime was spread on the arable fields to reduce the acidity level of the soil. This was not supplied in granular form but instead came as a very fine white powder. As a result lime was quite unpleasant to apply, particularly if it was even slightly windy. Supplied in paper sacks which we sliced open with a knife and emptied into the hopper of the drill, the lime had to be spread evenly along its length. It got everywhere, and within a short time you were white from head to toe as was the horse and the machine. The lime stung when it entered cuts or got into your eyes. I was perhaps lucky as my glasses gave some protection but within minutes of starting they were covered. Nonetheless I hate to think what it would have been like without them.

The concept of using protective clothing on the farm and having it supplied was still many years distant. Not until the Safety at Work legislation of 1974 did this really begin to change. Neither my employer nor George used anything other than weather protection and probably never recognised the need for it. The long-term history of farm labouring was a record of hard work carried out whatever the circumstances or discomfort. This was what was expected of the farm worker, male or female. I suspect that many of us even took a macho pride in being able to do the work, particularly the lifting and carrying. Few in this country could or would do it now, or would want to. The work ethic has changed, society has moved on and few would regret the changes. The bad backs and rheumatism suffered by older workers were the evidence and results of this arduous labour, but we younger ones took little note, instead delighting in our strength and ability. I was lucky. I only spent ten years on the farm and during my teenage years most of that was part-time. So I was spared these problems, except for back pains that are more likely to have been the result of an injury sustained at work when unfit after leaving university than from my period of regular farm work.

These days, although many soils still need liming and always will, the lime is handled in bulk and applied with high-capacity specialised spreaders, usually by a contractor. Although clouds of lime may be seen behind the machine the operator is well protected from it in a cab that is likely to be air-conditioned.

Despite these disadvantages spreading fertiliser was a job I enjoyed, even though it was tiring. Apart from the physical demands of loading the sacks and lifting them onto the spreader, I had to walk more than six miles to cover a six-acre (2.5 hectare) field and neither rough grass nor the soft soil of an arable field is easy to walk on.

One of the last occasions that I did the job almost brought disaster. While on one of my last vacations from university George sent me to spread fertiliser on a pasture field not far from the farm. The horse was a grey, yet another Prince, the very last horse in the stable before the farm converted entirely to tractors. I doubt if I drove him more than once or twice, for which the horse was probably very glad given what transpired. The field was not drained and was renowned for being very wet in parts, so much so that reeds grew in some areas. Although there was no obvious standing water I should have looked at these reedy areas prior to starting, to check how wet they were and then avoided any particularly bad spots. Instead, I set off driving backwards and forwards as usual and eventually through one of the reedy areas. To my horror, and no doubt also that of the horse, he sank almost up to his hocks. Fortunately, I was able to calm him and persuade him to back up onto firm ground. Had he gone down or continued to thrash around I don't know what the outcome would have been. I never told George or anyone else what happened. Fortunately, he didn't see us return to the farm after we had finished the field both covered in dried mud. Like the rest of the farm, this particular field was built on years ago. I trust that it was well drained before the houses were built.

Later in the spring, farmyard manure was spread on the fields in which potatoes or cow cabbage were to be grown. FYM is a mixture of the dung and urine from animals and the straw or other material used as bedding. Muck, as we called it, supplied some valuable nutrients, particularly trace elements, but its greatest value was in improving the soil structure. It was recognised that these two crops benefited most from a heavy application.

Some of the winter's accumulation of muck would already be heaped in the field. This would have been *drayed* (carted) out when the ground was dry, not a very frequent occurrence in North Devon in winter, or on the rare occasions when there had been a hard frost and the surface of the ground was frozen. Now the rest had to be carted out in the horse-drawn butts and all of it spread.

The muck cleaned out twice daily from the shippon was piled in loose heaps. The remainder was in the yards where cattle had been overwintered. These were several feet deep in muck and the bedding straw that was regularly spread in the yards for the animals to lie on and to absorb the urine so that the cattle kept reasonably clean and dry. This had been trodden in by the beasts but was only partially rotted and so the muck in the yards was quite hard to extricate.

We used long-handled forks with four curved prongs to prise it out and load into a butt. The long handles enabled us to exert considerable leverage to tear out the straw-bound material and the handles would bend with the stress if the fork was dug in too deep. Even if it was not, pulling out the material was a struggle, and often the mass that finally tore away was too bulky and heavy to lift at one go. Often, too, it was not balanced evenly on the fork and as the long handle had no T-piece it had to be gripped tightly to prevent it turning in one's hands. This could cause blisters until one's hands had hardened up. Forking the loosely piled heaps of muck from the shippon into a butt was easier and quicker.

Once the butt was piled high with muck the horse was led off to the field where the muck was to be spread. On arrival at the field the tailboard was unpinned and removed and the muck dragged off with the same dung fork used for loading. It was offloaded into heaps, three or four from a load evenly spaced in orderly rows up and down the field. When only enough for one heap remained the rest could be tipped out.

The body of the butt was pinned down by a cranked rod across the front that acted as a locking device. This had to be released before the body could be tipped. Each end of the rod fitted through a hole in a tongue that projected vertically from each shaft. To release the body the rod was rotated so that it could be slid sideways out of each hole in turn. The front of the body could be lifted off the tongues. To tip the body the horse was urged to back up to push up the front of the butt. When the remainder of the load had slid out, the horse was pulled forward to lower the cart body back down. Having re-slotted the cranked rod back into position, and remembered to go back to the first heap to collect the tailboard, it was time to return for the next load.

What followed was the best part of the job – riding home perched precariously on the front corner of the cart body, your feet positioned one behind the other on one shaft. Wisely, I would take an empty sack with me to put on the front edge of the cart body to protect my clothes from the muck on it and to serve as a cushion. However, nothing cushioned the jolts transmitted from the uneven rutted ground to the unsprung cart by its big iron-tyred wooden-spoked wheels. Once we reached smoother metalled roads the journey improved considerably but it was far from being the smooth ride we expect from a modern vehicle.

Once the field was covered in the heaps of muck they were spread, by hand of course. This was done with the same long-handled forks used for loading. The knack was to scoop up a fork load and swing it round, finishing with a twist of the wrist to spread the muck evenly in an arc over the ground, any lumps usually breaking up and scattering when they hit the ground. Once all the heaps had been spread the whole field was evenly covered with muck.

When the temperature started to rise and spring had really arrived we moved on to seed sowing. Most of the ploughing was completed in the autumn but before the crops could be sown the furrows had to be broken down to a satisfactory crumbly tilth. In general, the finer the seed the finer the tilth required in the surface layers into which the seed was sown.

A range of cultivating implements was used. A number of these have gone out of favour and been replaced by new types. Most of the ones we relied upon used tines – steel fingers or spikes that were dragged through the soil at various depths to break up the furrows and shatter any clods that formed. In some examples the effect was enhanced by using springy tines rather than rigid ones, or by having different shapes and sizes of 'coulter', or foot, at the bottom of each tine. Different implements were used in sequence to make use of their particular characteristics and effect upon the soil. Occasionally, rollers were also used in an attempt to crush hard clods that were resisting the impact of tines. These were solid wide rollers made in one piece of stone or concrete and were about 12–18 in. (30–45 cm) in diameter and mounted in a wooden frame.

However, the implement that we used most frequently at my home farm was the disc harrow. This worked on a different principle from the tined cultivators and consisted of four gangs of saucer-shaped steel discs fixed on axles in two rows in a vee formation, one behind the other. The front and rear gangs could be angled in opposite directions by means of a handle, and the greater the angle the greater the depth to which the discs penetrated. Discs were very effective on our soils and produced the desired tilth more quickly and with fewer passes than other implements. Nevertheless, we used tined cultivators when deep tilths were required for crops such as potatoes and for tasks like covering seed that had been sown, for which light zigzag harrows with short tines were used.

When I reached university I was surprised to hear the professor of agriculture totally condemn the use of disc harrows. His argument was that they forced a tilth by slicing up the clods and smearing the cut surfaces rather than by shattering them as tined cultivators did. He argued that this had an adverse affect on moisture exchange and that the discs cut up and so proliferated stoloniferous weeds like couch, or 'stroyl' as we called it in North Devon.

Stoloniferous weeds are those that can reproduce vegetatively by sending out subterranean stolons that root and produce a new plant at some distance from the parent. Even small pieces of stolon are capable of regrowth. We revered the professor who was universally recognised in the agricultural world as a great authority, but who was equally at home having a drink and chatting with students after an evening visit to a farm. However, on this one matter I am convinced that he greatly exaggerated the disadvantages. Subsequent research, which I witnessed whilst carrying out post-graduate studies, showed that couch and other stoloniferous weeds can actually be controlled by chopping them up, as they finally become exhausted if they are cut up again as soon as they begin regrowth. Certainly, large, heavy sets of discs matching the big tractors of today are still widely used.

Most of the cultivations were done with the tractor, but on my second farm we used a pair of horses to harrow seed beds after they had been sown to cover the seed. We also used horses to roll the seed bed prior to drilling fine seeds such as grass and to roll autumn-sown crops when the ground had become fluffy due to frost. The horses marked the soft seed beds much less than tractor wheels which tended to leave ruts. Modern tractors are fitted with wider tyres and double wheels or cage wheels to reduce this. This horse work at my second farm was usually allocated to me.

The two horses walked side by side, their inside bit rings connected together so that when I pulled on either of the rope reins tied to the outer ring of each horse's bit my command was also transmitted to his partner. Each horse was harnessed to a 'swingletree' by trace chains from the hooks on his hames. The swingletree was a short pole just wider than the horse with a hook at each end for the trace chains to be attached. The centre of each swingletree was linked to a common swingletree and in turn the centre of this one was linked to the harrows or to the frame of the roller. This arrangement allowed the pulling effort by each horse to be equalised and permitted some individual movement which was needed when turning corners. It also enabled me to see whether one horse was pulling harder than the other.

Making a U-turn at the end of the field with the harrows was reasonably straightforward, provided that the horses were not allowed to turn too sharply. Turning with the roller was not. The solid one-piece roller could not be turned through a tight turn without it slewing round. This resulted in a large wedge of soil being dug out at each end of the roller and damage to the crop. To avoid this a direct U-turn was not made. Instead, when the end of the field was reached the horses were turned gradually so that the roller also gradually turned until it was at an angle of about 45 degrees to the hedge. Then the horses were edged carefully back around the end of the stationary roller so that they pulled the wooden frame back over the roller. Then the horses were set off pulling the roller in the opposite direction, turning them gradually again to straighten the roller onto the new line. This technique required precise control of both horses throughout the turn. If the trace chains of the horse on the inside of the turn were allowed to go slack it was almost inevitable that the end of the swingletree would catch in the ground or the end of the frame of the roller and twist over, wrapping the trace chains around the horse's legs. Understandably, both horses objected to this, and they would show their displeasure kicking themselves free of the trace chains and by turning their heads around to give me a reproachful look. I learned that it was easier to let them get on with it and sort out the mess afterwards, rather than to try to stop them from kicking themselves loose.

I enjoyed working a pair although walking behind them for most of the day in soft soil was quite tiring. There was a companionship working with horses and an understanding and trust that developed between man and animal that is hard to describe. On a fine spring day it was the best of jobs. There was almost no motorised traffic on the lanes nearby and so the silence was only broken by the jingle of harness chains, my own occasional words of encouragement, the sound of larks overhead and birdsong from the hedgerows.

The only sound that marred those otherwise peaceful days was the blasting that took place regularly each day, just before lunch in the quarry at the other side of the valley, the sudden noise causing the horses to toss their heads and side step in surprise and fright. It reminded me, if I needed it, that it was almost time to stop for lunch and to rest the horses.

Regretfully, the volume of birdsong that I used to hear in the countryside is greatly diminished and the sight and sound of soaring larks is becoming a rarity. Sadly, my grandchildren will never have the pleasure and enjoyment of watching as many birds and being treated to the great variety of birdsong that I enjoyed. Changes in farming practices and cropping, the excessive use of pesticides and the widespread removal of hedgerows that formed the habitat and food source of so many of our bird species have decimated the population of field birds in most counties. Many species that I saw regularly, sometimes in great numbers, are now rare or are considered to be in danger.

Lunch was taken sitting at the edge of the field in the shelter of the hedge. I unhitched the trace chains from the horses and they moved over to the hedge to graze, munching away quietly apart from the slight musical jingle of their bridle chains. Invariably Carlo the farm dog who regularly accompanied me to the fields, and when I moved stock or fetched the cows home for milking, would appear from nowhere eager to share my lunch. He would have spent the morning rabbiting in the hedgerows, occasionally returning to accompany us for a while as we plodded backwards and forwards across the field.

Carlo, the sheepdog was a large black short-haired dog of uncertain parentage. There was only a small flock, which was just as well as Carlo was not a good sheepdog. He was too rough to herd sheep and was inclined to nip their heels if they did not move as fast as he thought they should. However, herding bullocks was a different matter. Many farm dogs are often afraid of cattle, particularly if the animals turn and defy the dog or even charge it. Carlo stood no nonsense. He dashed backwards and forwards behind the herd barking at any animal that hesitated and if that didn't suffice he would dart in, nip the heels of the offending animal and be away before the startled bullock had time to lash out at him. His party piece when dashing from one side of the group to the other was to grab the tail of the last animal and to swing round on it landing facing the other way ready to hurtle off in the opposite direction. We tried everything to break him of this habit but we were never totally successful. Fortunately, he had a very soft mouth and never to my knowledge damaged an animal or broke the skin.

As soon as a suitable tilth for the crop had been achieved the seed could be sown. Grain and root crop seeds were sown by drills similar in outward appearance to the drill used for spreading fertiliser. However, in the grain drill the seed was metered and fed down flexible tubes to individual coulters which delivered the seed right into the soil. Metering was done by a pair of small gear wheels, meshed together and driven by the drill wheels, which forced a measured quantity of seed into each coulter tube. The coulters were spaced 7 in. (18 cm) apart, in staggered rows across the width of the drill.

Our grain drill was a converted horse-drawn machine and was fitted with a foot board across the back. My job when a boy was to ride at the back of the corn drill standing on the foot board. From time to time I would raise the lid of the hopper and check the amount of seed left and ensure that what was left in it was spread evenly along the length of the hopper. When the hopper was almost empty I would shout to the driver in good time so that on the next occasion that we passed the seed sacks at the gate he would stop to refill.

The drilling was done backwards and forwards across the field, making a U-turn at each end. Just as when spreading fertiliser, the driver had to steer very accurately to ensure that each run matched the previous one and that there were no gaps or overlaps. The distance away from the previous coulter mark was difficult to judge. Modern drills are fitted with markers to overcome this problem, but our drill was a converted horse-drawn machine and it was probably easier to see the previous coulter mark when walking behind it driving the horse.

As we neared the end of a run across the field my other duty was to push down a handle that raised the coulters out of the ground and stopped the flow of grain. This prevented the coulters from being damaged when the tight U-turn was made at the hedge. Once the turn had been made I lowered the coulters back into the soil at the correct moment to start drilling again. The fields were seldom truly rectangular and as in almost all field operations a number of short runs had to be made to finish the field. The turning area, or 'headland', was drilled by going around the field several times after the rest of the field had been drilled. On modern drills designed to be pulled by a tractor the controls face forward so that no one is required to ride on them and drilling becomes a one-man operation.

On both the farms I worked on, all the root drilling

was a one-man operation done with a horse-drawn machine. Root drills were similar in design and operation to the grain drill but the coulters were spaced much further apart at 21 in. (54 cm) and only drilled four rows at a time. In our machine the seed was picked up and delivered into the coulter tubes by small cups mounted on a shaft driven by the land wheels.

Root drilling required even more accurate driving than grain drilling to ensure that no areas were missed or drilled twice, as any such errors were glaringly obvious once the crop had germinated. It was also a matter of pride that the drills, like ploughed furrows, were dead straight. All farmers look over the hedges as they travel about to see what their neighbours are doing and take great delight in noting any work that is less than perfect. Although I drilled grain with a tractor I was never entrusted with the more skilled root drilling operation with a horse.

Grass seed was not drilled; instead it was broadcast by hand. There was a device available to assist in doing this called a fiddle, presumably because it was operated by a bow rather like a violin bow. Seed was fed from a container onto a circular plate mounted on a spindle which was rotated by the bowstring wrapped around it, first in one direction and then in the other as the bow was pushed to and fro. Although I have seen a fiddle, and there was even one in the barn, I have never used one or even seen one in use. We simply took carefully judged handfuls of seed and flung them away with a sweep of the arm and a flick of the wrist to spread the seed in an arc. The knack was to harmonise walking speed with the movement of the arms so that the area spread with each throw matched the previous one to cover the ground uniformly. Care had to be taken to ensure that each handful was the same size to ensure an even spread. Skilled sowers used each hand in succession to widen the area that they covered on each traverse of the field. The seed was carried in a galvanised elongated tub shaped to fit one's body, carried at waist level and supported by a strap across the shoulders.

I was never considered to have the skill to graduate to this operation, although I did broadcast fertiliser using this technique and still do so on my own lawns. It was another tiring task if other than a small area had to be broadcast as the tub held a considerable quantity of fertiliser and the strap soon began to cut into the back and shoulders as you leaned back to counter the weight of the tub.

Chapter 11

Early Summer

As spring merged into summer there was a brief quiet period on the farm. The cattle were all lying out, with the milking cows brought in twice a day to be milked and then turned out again. The bullocks and sheep had to be visited each day to check that none had broken out and that they all appeared to be in good condition, with no foot problems or maggots. By this time all the older lambs were quite well grown and had left their mothers to gambol and play in groups, although at our appearance with a dog they would all dash back to the security of their mother's side. This would be accompanied by much bleating as each parent tried to find her offspring, and there were often two to locate as the native Devon Closewool breed produces many twins and occasionally triplets. Some things have not changed, and similar scenes can still be witnessed in every livestock area of the country.

However, there was some field work to be done. Root crops emerged in early summer and had to be singled and hoed to remove seedling weeds. As soon as the rows were discernible the steerage horse hoe was taken through the crop to hoe out the weeds between the rows. It was when carrying out this first hoeing that the horse had to be led because it couldn't be relied upon to identify the rows of tiny seedlings and to walk between the drills. When the plants were larger and the horse could see the rows clearly, leading became unnecessary. Then it became a one-man operation, with the horse allowed to follow the rows with loose reins once it had been turned into each row. Sometimes, even I found it difficult to see the drills at first, particularly when germination had been variable and some parts of the row had still to emerge.

The horse hoe spanned several drills, ideally the same number that were drilled at one pass so that any deviations from the straight when drilling could be followed exactly when hoeing. The hoe was fitted with three blades for each inter-row space, a triangular one that ran down the centre and L-shaped blades that ran down each side. The positions of the L-blades were adjusted until they ran as close to the plants as possible without damaging them. To be able to follow the rows accurately without wandering into the plants the hoe frame was fitted with wheels and a steering device to guide the hoe along the drill, which was operated by the worker walking behind.

On my second farm we used a steerage hoe mounted on the hydraulic lift system of the tractor. This was operated in exactly the same way as the horse-drawn hoe except that the worker at the back steering the hoe had a seat and did not have to walk.

Once the first hoeing had been completed and the seedlings were sufficiently well grown to stand it, the crop was singled. Only mangolds required this because the mangold seed that we sowed was really a fruit containing more than one seed and so two or more seedlings often germinated close together, sometimes intertwined. We had to walk along each row armed with a draw hoe, a long handle with a transverse blade mounted on a swan neck, and carefully separate out and remove all but the best plants from each cluster. At the same time we would gap the plants so that they were spaced at about 12 in. (30 cm) intervals, and remove any weeds between the plants that had been selected and retained. If the row was thin in places due to poor germination more plants would be left next to gaps to compensate. When I reached university I learned that research had established that about 25,000 plants per acre (62,000 plants per hectare) was the optimum number to maximise the crop. The row width and spacing within the row that we used in common with most other farmers gave approximately that plant population. As this was relatively recent research I realised that, like so many other practices in farming, traditional plant spacing had been established as a result of farmer observation and trial and error over many generations.

Singling was a slow, tedious task that had to be done with care to avoid damaging the seedlings that were to be retained. Plant breeding has since eliminated the need to single by producing single germ seed that is pelleted and of uniform size so that individual seeds can be planted at the required distance apart by spacing drills. Coupled with the use of pre-emergence weed killers, plant breeding has also obviated the need for hand hoeing.

Other root crops such as swedes were naturally monogerm and so they did not require singling, although they still had to be hand hoed to gap the plants and to remove weeds from within the rows. The seed drill sowed the seed in a steady stream to allow for indifferent germination. As a result there were more plants than required in most parts of the fields, and the surplus was hoed out. If significant weed growth recurred within a root crop before they grew large enough to smother it, the horse hoeing and hand hoeing had to be repeated, although hand hoeing was much more rapid as there was no need to gap or single.

Hand hoeing in other parts of the country was sometimes done as piecework by itinerant gangs of workers or by the farm staff in the evenings. I do not recall this happening on my two farms in North Devon. All our hoeing was done during the normal working day or during the evenings as overtime. However, farm staff were expected to hand hoe any rows of potatoes that they had been allowed to plant in the field, in their own time.

Swedes and turnips were drilled during the early summer; traditionally we drilled swedes on midsummer's day.

Other important work carried out during early summer included sheep dipping and shearing, and haymaking. As the activities normally took place during the school term, I was only able to take part on the rare occasions when they happened at weekends. So my experience was largely confined to weekends, my full-time employment after leaving school, and the long university vacations that extended from early summer through to autumn. It is as well that vacations lasted as long as they did because, without my wages and the massive amounts of overtime that I did, managing on my county major scholarship would have been difficult.

Haymaking was still the principal means of conserving grass to feed to cattle, sheep and the horses during the winter months when there was little outside grazing to be had and the cattle had to be housed to prevent poaching of the land. Very soon after I completed my practical experience, silage replaced haymaking on the majority of farms up and down the country for feeding cattle. Grass to be ensiled is cut much earlier in the year than hay when it is higher in protein and more nutritious, resulting in a more valuable feed, if it is made properly. As it is cut earlier two cuts or more are possible and as the process of ensilage is more akin to pickling than drying the crop has to be wilted only briefly before it can be transported to the silage pit.

Hay cannot be baled or stacked until the moisture has been dried out of it. This takes longer, particularly if there is rain during the period that it is lying in the field, and prolonged exposure to rain and sun reduces its feeding value. Excessive bleaching by the sun reduces the carotin content.

However, as hay was not cut until the end of June or even July, depending on weather and when a field was 'laid up', that is closed to stock, ground-nesting birds had the opportunity to raise their young in peace. Undoubtedly, the changeover to silage, coupled with the widespread and excessive use of herbicides, has led to serious reductions in the numbers of such species as larks, lapwings and wild game birds, and the almost complete disappearance of other species like the stone curlew. In addition, there has been a reduction in the numbers of small mammals that once lived and nested undisturbed in hayfields.

Grass for hay was cut with a 'grass machine', or mower. Two horses were required to pull it. A sturdy pole extended from the front of the machine, and the two ends of a crosspiece linked to its end were strapped to the collars of the horses to support the mower. Trace chains hooked to the hames were led back to a conventional swingletree system like that used when harrowing, and on all occasions when horses were in traces rather than between shafts. Likewise, when horses worked as a pair the inner bit rings were linked together and the driver controlled the horses with rope reins tied to the outer rings of the bits. The driver sat on a shaped cast-iron seat projecting low down at the rear of the machine from where he could also reach the long handle used to raise and lower the cutterbar at each corner of the field. The cutterbar was a reciprocating blade assembly almost identical to the one used on the binder. The blade was driven to and fro by a connecting rod from a flywheel driven through a gearbox by the two iron wheels which supported the machine. The rims of the wheels were ribbed to prevent them from slipping on wet grass. A 'swath board' at the outer end of the cutterbar swept cut grass aside so that on the next circuit of the field it would not get caught up at the driving end, blocking the knife and preventing it from cutting.

Regular sharpening of the blade was even more essential than when cutting corn. Blunt blades did not cut the grass cleanly and led to blockages as well. They also increased the draught of the machine, tiring the horses more quickly.

I never had the opportunity to drive this rig. By the time I was able to take part in haymaking at my home farm during my first summer vacation from university, the horse pole had been shortened and fitted with a hitch that enabled the mower to be towed by a tractor. However, it was still necessary to have a rider on the mower seat to raise and lower the cutterbar so it was questionable whether much progress had been made, although unlike horses the tractor never tired. At my second farm the mower was designed to be pulled by a tractor instead of being a conversion and the controls could all be reached from the tractor seat. The mowers we used, like those I saw at neighbouring farms, were manufactured by either Bamfords of Uttoxeter or by Bamletts of Thirsk. Their design and appearance was remarkably similar in my eyes and I thought it curious that the makers' names were also so similar.

I also used the tractor mower for cutting weed in grass fields, usually clumps of thistle. The infestation seldom justified cutting the whole field but doing this job provided an excellent opportunity to search the field for mushrooms as I drove around. They regularly grew in certain permanent pasture fields and were particularly common in the fields where the horses regularly grazed.

Once the grass had been cut for hay it was necessary to fluff it up and spread it about as rapidly as possible so that the wind could dry it. This was done with a tedder. A number of toothed rake bars were mounted on a cylindrical framework rotated by the land wheels of the tedder. As this cylinder turned, the projecting teeth picked up the grass and carried it up and back over, tossing it out from the rear so that it floated gently down to the ground to form a loose 'windrow'. A curved sheet metal guard behind the driver's seat deflected the grass rearwards and protected the driver from falling backwards into the moving parts. The tines had to be retracted before it could be moved out of the field. Only one horse was required to pull the machine and it was a very pleasant job on a fine and warm summer's day.

Tedding was repeated a day or so later to increase the rate of drying further. If heavy rain fell on the grass, flattening it to the ground, it would have to be tedded again. However, too frequent tedding had to be avoided, particularly when the crop was nearly dry, because leaf was easily broken off and lost. The thin leaves dried out more quickly than the thick stems and are nutritionally more valuable, so their loss was serious.

The quality of the hay depended on the grass species, the stage of growth of the crop and how well the hay was made. Many of the fields were meadows or permanent pastures. They contained many different grasses and clovers as well as plant species that many farmers regarded as weeds. Many of these flowered throughout the spring and summer and resulted in the lovely flower meadows that were captured in generations of landscape paintings and in the words of our romantic poets. Alas, few such flower meadows still exist. They have all been ploughed out and replanted, or the weeds and flowers eradicated with weedkillers to increase the yield of grass. However, at university it was argued by some that animals thrived better on established grassland because the range of species made the crop more palatable to the stock and supplied a wider range of valuable nutrients and trace elements.

The rest of the grass fields were 'leys' which were ploughed out after a few years as part of a rotation that included other crops, or they were directly re-sown with a specialised grass seed mixture intended to give greater production.

The best hay was made very quickly so that the nutrients were not bleached out by the sun or leached out by rain. Weather played an important part in hay-making. Ideally, the grass would be cut when the weather was set fair for several days. I cannot recall whether there were local weather reports on the radio then. Certainly the farmers that I worked for relied on weather lore and their interpretation of barometer readings to decide when to start mowing, but it was not easy to predict periods when it would be dry for several days by those means. Even the present-day meteorologists with their worldwide data collection capacity and computing power do not seem to be able to do that with any degree of reliability. In some years there were no such periods and in those catchy conditions the hay had to be made as and when breaks in the weather allowed. As a result some of the hay made in difficult years was very poor and of low feeding value.

Once the crop was considered to be dry, it was ready to be carried and made into a 'hay rick', that is a hay-stack, or baled. If it was slightly damp it was *doaney*, and it was unwise to carry.

The first step to save the hay was to rake it up into rows. A horse-drawn rake was used to do this. We used a Huxtable patent expanding rake manufactured by a long-established Barnstaple firm. This machine was narrow enough to travel through the tight North Devon lanes and gateways but once inside the field could be expanded to rake a greater width. When the rake arrived in the hayfield a leg at each side of the machine was released and the horse moved forward to lift the rake off its wheels and onto the legs. The rake was then expanded sideways by cranking a handle. The long, curved rake tines which were bunched together in the transport position were connected by a clever linkage system that ensured that they spread out evenly as the machine expanded and widened. Once extended the machine was backed to lower it down onto its wheels and the legs were clipped back up out of the way.

The rake was driven up and down the field, with the collected hay regularly dumped in windrows across the field by pulling the lever that raised the rake tines, thus releasing the bunched up hay.

If the hay was to be carried home and put into hay ricks, a Dutch barn or up into the stable tallet, it was forked straight into the hay carts – the same carts used to transport sheaves of corn at harvest. Pitching hay was much more difficult and tiring than loading sheaves because the hay had been rolled up by the raking operation and it was not easy to pull out a small enough amount from the tangled roll to be able to pitch it up into the cart. More often when you dug in the pick fork and heaved, a huge mass began to move and you had to try to tease out a smaller quantity.

Pitching hay to a trailer in a field was the only farm job that beat me, and then only once. It occurred on my second farm on a very hot day; the hay was dry and dusty, and after battling for several hours I became too exhausted to continue. My legs and arms became weak and shaky and I had to sit down and rest for some time before I could continue. I now suspect that dehydration was the main reason for my virtual collapse but we were all unaware then of the problems that this could cause.

Pitching off the carts was also hard but not as hard as pitching in the field, particularly if you had made the load and knew the sequence in which the layers of hay had been arranged. Even so, it was quite common to find that you were standing on part of the fork load that you were trying to extricate. When making the load the person on the cart guided or moved each fork load into the right position as it was pitched up, working around the perimeter of the cart and then filling in the middle, overlapping the outer layers to secure them. It was necessary to keep moving and treading to compress the hay to make the load more stable and to maximise the size of the load. Once loaded it was roped. Then the load maker could dismount by lowering himself down the front of the load, stepping first onto the front lade, then placing one foot onto the horse's back and the second down onto the shaft before jumping down to the ground. It was exactly the same procedure used for dismounting from a load of corn.

Instead of carting it home the hay was often stacked in the field. This speeded up the operation, and once it was finished the hay rick was thatched and left in the field until the autumn or winter when the hay was required to feed the stock.

The hay was moved to the site where the stack was

to be made with a hay sweep. This consisted of a number of long, pointed, wooden tines fixed side by side to a crosspiece with a vertical back that looked rather like a hurdle. The sweep was pulled by a horse using double-length draught chains connected to a swivel at each end of the crossbar. In operation the sweep was pulled along a windrow that had been raked up, the tines sliding along the ground under it and sweeping up the hay. When the sweep was loaded it was turned out of the windrow and dragged to the spot where the rick was being built. The driver, who walked behind, unloaded it by pushing up on the two handles fixed to the back of the sweep so that the points of the tines dug into the ground. This caused the sweep to rear up and roll over the load of hay, leaving it in a heap as the horse moved on. The swivels to which the trace chains were attached allowed the sweep to make a complete revolution, finishing up with the tines on the ground again ready to sweep up another load. It sounds an easy job, as the horse knew what to do and needed little guidance. But the driver had to be constantly alert because if there was an irregularity in the ground the tines tended to dig in and unless the horse was stopped immediately the sweep would rear up and roll over the hay that had been collected. Then there was a struggle to get the sweep back into position to collect the hay that had been dumped prematurely.

On my second farm we used a sweep with much longer tines mounted at the front of the Fordson Major tractor. It could be raised when loaded, making it easier to travel to the site of the rick as the tractor was almost impossible to steer with the loaded tines rubbing along on the ground. The sweep was raised by steel wires linked to the top of the back frame and routed back along the sides of the tractor and across the floor plates to be finally connected to the hydraulic lift arms. For this operation the normal lift rods and links were removed and the upper lift arms turned vertically downwards to allow the cables to be attached with pins through eyes at the ends of the cables. This worked quite well and I became quite adept at operating it. However, great care and judgement were required when turning or manoeuvring close to a hedge. Any steering change was greatly exaggerated by the long tines stretching far out in front of the tractor, and it was very easy to catch the tines in the hedge as the tractor swung around.

However, I discovered that the tractor/sweep combination had one very nasty and dangerous trick in store for me. The ends of the steel wires were doubled over and held together with bolted metal clamps to form the connection eyes. Whenever the sweep was lowered, the cables and metal clamps whistled forwards across the floor plates of the tractor. One afternoon when I lowered the sweep the clamp caught the heel of my boot and wedged it against the clutch pedal, forcing it into the ‘down’ position and jamming it there. The hydraulics on the original Fordson Major were driven by the transmission and would only work with the clutch engaged and the clutch pedal ‘up’, so I was trapped. The only remedy was to switch off the engine, unlace my boot, take out my foot, limp around to the front of the tractor and struggle to heave up the tines of the sweep to slacken the cable and release my boot.

This I managed to do as the sweep was not loaded, and when I raised the tines and the wire slackened my boot fell free. A lump chewed out of the heel of my boot helped to remind me to keep my heel well clear of the wires in the future. Of course, an arrangement with exposed wires and metal clamps would not be permitted under today's safety regulations. Fortunately, the incident occurred at the far end of the field as I lowered the sweep to begin loading, and so no one saw what happened and my blushes were spared.

At my home farm the hay had to be pitched up to the rick, as we had no type of elevator. This presented no problem until the rick was about 8–10 ft (2.5–3 m) in height and the top was beyond the reach of the pitcher. Even to this height pitching was not easy. The action of the sweep as it swept up the hay and then dumped it tended to roll up the hay, and made it even more difficult to disentangle fork loads small enough to be lifted comfortably and pitched up onto the rick than when loading from the windrow. In order to build the rick higher and to finish it off with a ridge, a cart was dropped beside the rick with its shafts on the ground as a halfway stage. From then on the hay was double pitched, one worker pitching from the ground onto the cart and a second standing in the cart forking the hay up onto the rick. As the rick got higher, hay was allowed to build up in the bottom of the cart to raise the level of the second pitcher.

At my second farm this laborious and labour-intensive method for getting the hay onto the rick was avoided by using a hay pole. This consisted of a long thick pole like a telegraph pole, erected at the rick site and held in a near-vertical position with guy ropes. A rope with a grab connected at one end was threaded over a pulley attached to the top of the pole and the other end hitched to a swingletree and trace chains, so that a horse could be harnessed to raise the grab. Instead of pitching, the loader dug the tines of the grab into the pile of hay which had been swept in and then led the horse away to raise the grab. When it was at the required height the grab was swung over the rick and opened to dump the load of hay. This worked well provided that the loader grabbed reasonable amounts but if the grab-load was too large it inundated those working on the rick and made it impossible to spread the hay around evenly and build a stable rick.

On one occasion this happened when another young worker, a part-timer, decided to 'wind up' the rick maker and sent up the hay as fast as he could. The rick maker was on his own and before he had spread one massive load the next had been dumped on it. More and more hay was being lifted each time, and I was having difficulty in bringing in enough hay with the tractor sweep to keep up with him. Finally, a load so big was picked up that the horse was pulled back onto its hind legs when the draught increased as the grab swung in over the rick. This sobered the loader and the work proceeded more sedately afterwards. However, there were repercussions as the rick had to be posted with forked tree branches on the two long sides to support it and to prevent it from toppling over as soon as it was completed. My employer was not in the best of humours that night. He was even less pleased when the hay in the rick started to get hot. It had been gathered on a Saturday to avoid carrying it in on the Sunday and it wasn't really dry enough.

There was always a tendency to stack hay too soon and risk it overheating. Every farmer was keen to get his hay safely into the rick, especially if rain appeared imminent. However, my employer would never carry out any harvesting on a Sunday. He was a fervent member of the local Plymouth Brethren church and would only carry out essential routine work on the Holy Day. On another occasion, on a Saturday when it looked as if the hay would be fit to carry on the following day, I asked him if he wanted me to come into work. He stared at me and said, 'I'd rather see it rot on the ground than carry it on a Sunday.' I didn't offer again.

Occasionally, ricks that had been made before the hay was thoroughly dry became so hot that spontaneous combustion occurred. This took place at the centre of the rick and so the fire often had a strong hold before it became obvious, and then there was little chance of saving it. Our rick smelt hot, and when we tested it by pushing a thermometer probe – a long tubular rod containing a thermometer inserted near its tip – deep into the rick our fears were realised. It was hot, very hot. My employer decided that we would cut into the rick and extract the hot core. We worked together using hay knives to cut out blocks of hay to reach the hot area. Hay knives have triangular blades about 3 ft (1 m) long with one sharp edge and an offset tee handle parallel to the blade. The knife is pushed into the hay repeatedly, rather like using a saw vertically but pushing the tee handle with both hands. It was hard, hot work and it got hotter still as we got closer to the seat of the trouble. There the hay had turned brown, almost black in places, and it was clear that we had caught it just in time. The hay was caramelised and its feeding value destroyed, although cattle do find such hay palatable even if it does have no nutritional value.

We were black from head to foot and drenched with sweat by the time that we finished. I have since realised that what we did was quite dangerous. There was a risk that if the material was hot enough when we reached it any draught or gust of wind would have caused it to burst into flame spontaneously. As we were working in the deep hole that we had cut into the rick we might easily have been engulfed in flames before we could escape.

There was considerable skill in building both hay ricks and corn ricks. Although I have seen circular ricks in other parts of the country, I cannot recall seeing them in North Devon, and all our ricks were rectangular with ridge tops, hipped at each end like a house. Like corn ricks, the sides of hay ricks were never vertical but were allowed to slope outwards from the base so that rainwater running off the top dripped off the eaves rather than running down the sides and seeping into the hay. It was quite difficult for the builder to achieve the same uniform slope on all four sides of the rick. From the top of the rick it was not possible to see the inward-sloping sides, and so the rick maker had to descend from the rick from time to time to see how it was shaping.

I never built a rick, although many times I helped, on top, by moving hay over to the maker or by passing the sheaves to him on a corn rick when he was working at the far side of the rick. The rick maker worked steadily round the rick pushing hay or placing sheaves out to the edge, taking particular care at the corners. He continued working round and round, filling in the middle, overlapping each layer as he went. 'Keep treading boy' was his constant exhortation to me, as the helper's second role was to keep treading the hay to consolidate it.

Once the eaves had been reached the helper had a new role. A tiny space was left for the helper to stand in so that he or she could pass the hay up to the builder with a pick as the ridge was built and the rick completed. The ricks looked beautiful when complete, although like corn stacks a tendency to lean too far one way or the other had to be counteracted by pushing in a long, forked pole to support it. Once the rick had settled it was often possible to remove these. The hay rick built with the help of the hay pole was an unfortunate exception.

Once harvest had been completed, the rickyard at my home farm was a splendid sight with half a dozen or more corn stacks, all standing four square and neatly thatched, two Dutch barns full of hay and maybe a rick or two of hay also thatched. The rick maker had worked on the farm all his life and had learnt his skills from his father who had also worked there, and he took a great pride in his ricks, that he also thatched.

In my last two years of farm work, hay was baled and not carted loose. We used an International B45 baler that was driven by an articulated shaft from the power take-off shaft at the rear of the tractor. By then we had progressed to the new Fordson Major with the diesel engine. The earlier Standard Fordson had no power shaft, although it was possible to buy a baler of the same make and model which was driven by an engine mounted on it. The machine compressed the hay into bales that were 12 in. thick, 18 in. wide and about 36 in. long (approximately 30 × 45 × 90 cm), although the length could be varied. As the length was about twice the width, load making and stack building were simplified because the bales could be overlapped to tie them in to make a secure load or stack, just like courses of bricks in a wall.

At the same time when the baler was introduced the tedder and rake were replaced by a combined haymaker. This incorporated two rotors that could be adjusted to rotate in either direction and were fitted with curved tines. The machine could be set up so that it turned the hay or spread it for drying, but if the rotors were set to turn in unison they side raked the hay into windrows which were narrow enough to be picked up by the baler. It was argued that the turning action was gentler than that of the tedder. However, I doubt if I was alone in believing that the tedder left the crop fluffed up much more, so that it dried more quickly and that mechanical damage to the crop was no greater, provided that the tedder was not used when the hay was almost dry. I suspect that this proviso was often forgotten by farmers desperate to dry their hay in adverse weather conditions, and so the simple and effective tedder was denigrated and ultimately superseded. However, side raking was much quicker than with the rake as the action was more effective if carried out at high speed. Also, as it was towed round and round the field it produced a single continuous windrow ideal for the baler to follow and pick up.

The baler was also used to bale the straw after corn crops had been combined. It differed little from the medium-density baler still in use today, although the majority of balers now sold are 'big' balers which produce the large round bales or increasingly large square ones that are to be seen scattered around cornfields and stacked in farmyards.

As the baler moved along, rotating pick-up tines mounted on a cylinder picked up the hay and fed it into the bale chamber where a ram compressed it against the previous bale that was still held in the baler. The ram was a piston, driven backwards and forwards along the chamber by a crank and connecting rod. A toothed wheel pressing on the side of the bale being produced metered its length, and when it reached the pre-set length the metering wheel triggered two needles that passed two bands of baler twine around the bale and fed them into two knotters. The knotters were heavy-duty versions of those used on the binder; baler twine is much thicker and stronger than binder twine. Once tied the bales were forced along a slightly tapering chute and out of the baler by successive strokes of the ram as more material was forced into the front of the chamber and compressed to form the next bale.

Hay bales weighed about 40 lb (18 kg), straw bales less. However, if the material being baled was wet, or if the bales were packed too tight, they weighed a great deal more. We used to pitch them up onto the trailer with pitch forks, two of us working together to put up the heavy ones. Heavy ones could be lifted on your own if you did it in two stages. The knack was to dig the fork into the bale and then pull up with the left hand and push down with the right, grounding the fork handle to act as a fulcrum. Then the pick could be forced upright with the bale balanced on top. The final stage was to bend at the knees, grasp the fork low down with both hands and, with the fork handle steadied in the curve of the shoulder, lift the fork vertically to the height of the load.

Although we never made grass silage at my home farm, we did make arable silage during my last long summer vacation from university. This was a mixed crop planted in the spring with the intention of cutting it once and ensiling it. What matured was a tall, thick, tangled crop. When we attempted to cut it the mower blocked up every few yards regardless of how sharp the blade was, and we had to stop, get off the tractor and mower and remove the mass of material that had wound around the cutterbar before continuing.

If cutting was a nightmare, carting it was even worse. George had purchased a loader that was attached immediately behind the tractor. The loader picked up the crop and then conveyed it up an elevator which discharged it into a trailer towed behind the elevator. Unfortunately, the discharge height from the elevator was lower than the hay lade at the front of the trailer and so it had to be replaced with a sawn-off lade to allow the material carried up the elevator to flow into the trailer. The material was so thick and heavy that it flooded into the trailer in a continuous tangled mass. It was a Herculean task for the man on the load to drag it back with a fork to make some semblance of a load and to clear it from under the elevator. The result was a tail-heavy load. Before we dared unhitch the trailer much of the load had to be redistributed back to the front of the trailer again. The first trailers introduced onto the farm had a central axle just like the carts that they superseded and so, if we had tried to unhitch with a tail-heavy load the trailer would have tipped backwards.

Once we got home the dreadful stuff had to be pitched off into the silage tower we had erected. The circular tower was constructed of shaped pre-cast concrete sections bolted together, and was low in height but wide in diameter. Each trailer load was a tangled mass all twisted together and so, when attempting to separate a pickful, half the load seemed to move, and it invariably included what you were standing on.

The art of making good silage is to remove the air as anaerobic conditions favour the bacteria that produce the lactic acid that preserves the silage and conserves its nutritional value. We walked around endlessly on the wretched stuff in our silo but never managed to consolidate it adequately and to remove enough of the air trapped in it. Fortunately, I had moved on when the time came in the winter to feed what was in the tower. I never heard how they extracted it or what condition it was in. There was a curious hush when I asked about it. I suspect that it was transferred straight to the manure heap without the intermediate stage of being fed to the stock!

In between times during this busy period we 'pared', that is cut the roadside banks which were the farm's responsibility and any field hedges that might interfere with harvesting machinery. We used a grass, or 'paring' hook, with a long, almost semi-circular curved blade (called a sickle in other parts of the country). With this we trimmed the grass using a stick cut from the hedge in the left hand to clear the grass from the blade as it was cut. Like the scythe the hook was sharpened with a stone.

The trimmings were picked up with a horse and cart. This was another job for which a horse was well suited as it would move on when told, allowing the loader to continue working without pausing. Continually climbing on and off a tractor to move a trailer forward is tiring and time consuming.

A 'length man' did most of the roadside and verge cleaning and trimming in those days. Our local man was employed by the county council and lived in a tiny cottage some miles from his nearest neighbour. Known as the Tin Hut the house was both walled and roofed in corrugated iron. He travelled around the area for which he was responsible on a bicycle on which he also carried his tools, a broom, a long-handled Devon shovel, a long-handled fork and a hook, augmented with others as necessary. As well as trimming grass he cleaned out the ditches and ensured that the gulleys which drained water from the road surface into the ditches were kept clear. Sadly, local authorities dispensed with length men to save money long ago, more is the pity. Today, the silted-up roadside ditches that are filled with weeds and regularly overflow across rural roads, and the pools of water that result and are slow to run off because of blocked or filled-in gulleys are evidence of the service that those poorly paid but invaluable workers provided.

Chapter 12

The Milk Round

THE milk produced at my home farm was retailed and delivered door to door, every day, 365 days a year. The deliveries were made by George, his young wife and his father, although his father left the round each day at mid-morning to go to town to attend the market or meet his cronies. Not long after I had wandered into the cornfield for the first time, I began to go out to meet them and to accompany them on the round. They started at 7.30 am but I soon learned the route they followed each day and was able to find them later in the morning. It wasn't long before I was meeting George and his wife every Saturday morning and every morning during school holidays.

Initially, I helped by running back to customers when they needed something extra from their normal order. Apart from milk, which was still rationed at the end of the war and for some time after, potatoes, apples, pears, plums and tomatoes were all sold in season. For a favoured few there were large purple figs picked from a tree against a wall in the garden behind the farmhouse. Soon I was delivering milk, too, and before long I had my own customers who I went to whenever I was on the round.

The milk, fruit and vegetables were carried in a horse and trap. The milk was contained in two 10 gallon (45 litre) churns secured with straps at the rear of the trap. Each churn was fitted with a brass tap, and the milk was drawn off into pint or half-pint measures and then poured into carrying cans. One or two of the measures had no lips around the rim and it needed a steady hand to fill them exactly to the brim. More milk was carried in extra churns and when the churns with taps were empty they were lifted down onto the ground and the milk in the extra churns emptied into them.

The customer took delivery in different ways that had to be learned. Many customers would answer the knock on the door with a jug for the milk. Others left the jug with a cover over it outside, or in the porch. Yet others left the back door open and you walked in, called a greeting, and poured the milk into a jug that would be waiting on the kitchen table, the window ledge or some other convenient surface.

Two or three customers always had a hot drink ready and the member of the team who served them would disappear for several minutes, catching up further down the road. Inevitably, if one or other of the team was delayed, perhaps having to return with vegetables or change on a pay day, there were some variations in who we went to, but it was accepted that one or two customers were the private preserve of one or other member of the team. George's wife had several such customers and was particularly friendly with one young housewife, and always went to her house for a drink and a chat. Only once did I go there, when George's wife was missing from the round for some reason. To my surprise, when the door was opened the young woman was dressed only in a nightdress that left little to my imagination. I am not sure who was the most surprised. Being a rather naive teenager at the end of the 1940s, I suspect that it was probably me. I don't recall being offered the cup of coffee that was probably waiting either.

A few customers like my mother paid for their milk daily, but most paid weekly and a few were invoiced. A very small number paid erratically and sometimes had to be asked for the money they owed. On a Saturday I collected the money too. Many of the customers knew exactly how much they owed and had the correct change ready. I hadn't being helping on the round long before I was allowed to carry the money that I collected in my own pocket so that if change was required I was able to give it without having to go back to the trap for it. When I look back I am amazed and gratified that I was trusted so soon, as the quite large sum of money that I handed in when we returned to the farm was never checked.

As rationing was still in force when I started, customers were allocated to one supplier and as a result we supplied every house in the area covered by the round. It was largely a suburban district that we supplied and many of the houses had long drives. We must have walked several miles each day. Progress was much faster when we reached several streets of terraced houses that opened directly onto the road. The people in these narrow streets were exceptionally friendly, and almost every house had the open-door arrangement where we were expected to walk straight in.

The milk ration was quite small when I started. Several elderly ladies on the round who lived on their own were only entitled to a quarter of a pint most days, which cost one penny. It was a pitifully small amount and they were left more if we had it. Typically, they were more concerned that there wouldn't be enough for their cats than they were for themselves.

We did not always have enough milk to complete the round and had to call and tell every customer still to be supplied that we would deliver after the afternoon milking. Each year, during the period before the cows had started to calve, the output from the herd was insufficient to supply all our customers, and North Devon Dairies made up the shortfall. This led to complaints from some customers who insisted on having our milk because it had a much bigger head of cream. This led to a few arguments, particularly at the end of the round when there was very little milk left. The dairy's milk was pasteurised and the process homogenised the milk so that there was virtually no visible layer of cream. The dairy was accused of skimming it off before it was delivered! Despite these problems we retained most of our customers when rationing ended and customers were able to choose their supplier once more.

The trap was pulled by a horse called Punch. He was smaller than the other horses used for field work but larger than the usual trap horse. At busy times in the year on the farm he worked in the fields with the others. He was placid, quite unmoved by any traffic that passed and knew the round as well as anyone. He walked on from house to house with little or no instruction, and knew exactly where to stop after each move.

When Punch was retired some years after I started, a smaller mare called Trixie replaced him. She had an aversion to buses. This was unfortunate because part of our route was on a main road with a frequent bus service. She had several run-ins with them. Once she appeared to be attempting to climb onto the rear platform of a double-decker when it pulled into a stop just in front of her. On another occasion when a bus passed close to her she set off at a gallop, turning up an unpaved side road with the trap bouncing along behind. By the time she had stopped and we had caught up with her, one of the tap churns lay on the ground and the second hung from the back of the trap suspended by its tap. One of the spare churns had tipped over and the lid had come off. There was milk everywhere. Other similar incidents occurred when I was not on the round which I heard about.

After the war ended, it was not a great surprise when George decided to make Trixie redundant and sold her. To take her place he purchased a war surplus vehicle, the first motor vehicle on the farm, apart from the tractor. It was an ex-RAF Standard 10 pick-up with a canvas tilt and canvas roof over the cab area. George – who was a skilled carpenter and carried out many repairs on the farm, even making the wooden wheelbarrows that we used – replaced the canvas with a metal roof constructed on a wooden subframe and painted green.

This vehicle served us well for a number of years, although it did have its idiosyncrasies. The handbrake was cable operated and the cable stretched after a time so that it was not always effective. The round included houses beside a steep main road and it was normal practice to stop at a slight angle to the pavement, so that if the handbrake didn't hold and the van ran back the kerb would stop it. I was driving during one of my vacations and parked at a rather obvious wide angle. A passing policeman in a patrol car stopped and asked whether the brakes were faulty. Not waiting for my hesitant reassurance he climbed in, released and then re-applied the handbrake. Fortunately, the rear wheel hit the kerb at the same moment and the van stopped instantly. This satisfied him as he didn't seem to realise what had happened and he reluctantly drove off after admonishing me for parking so poorly. We breathed again.

This van was replaced first by a forward-control Morris and then by a yellow Commer which had a four-speed gear change on the steering column. Others followed it and the round was expanded, but by then I had left for good.

At about the time the horse and trap were superseded, the milk cans and churns were also discarded and replaced by bottles. The tops were cardboard discs printed with the name of the farm and pressed into a recess round the neck of the bottle. There were four different sizes of bottle, ½ pint, 1 pint, 2 pint and, unusually, 1½ pint. The introduction of bottles sharply reduced the time taken to complete the round. We did not have to measure out the milk for each customer and were able to carry enough bottles in wire carrier baskets to supply several customers without having to return to the van. However, the old intimacy with customers was lost as we no longer saw them except for on pay day. Now the bottles were left outside the door, notes were left for us, and sometimes the money, too, so we didn't see those customers even on Saturdays.

Although bottles saved us time on the round, the boring chore of bottle washing faced us when we returned. At first this was done by hand in a big sink with the diligent application of bottle brushes. Ultimately, the first bottles were replaced with ones similar to those still in general use today with foil tops and with the name of the farm on the side in red. The 1½-pint size disappeared, and the process of filling and cleaning was improved and mechanised.

By the time I reached my early twenties, I knew the round so well that when I returned home from university for a vacation I was able to slot right back into it. The changes since my previous vacation were usually small and were soon learnt. In any case I could consult the round books which were now in regular use. There were other changes though. George was doing the round with a young female employee. As he was now running the farm it made sense for me to do the round in his place, as it enabled him to be at home where his time was used more valuably. So she and I did the round, and we shared the driving. When we returned I moved onto farm work and she did the bottle washing before she finished for the day. By this time we were leaving the farm at 7 am and so each morning I walked the few yards from my home in time to load the van and have a cup of tea in the farmhouse before setting out.

In recalling my experience of the milk round I realise how much I enjoyed doing it. It brought me into contact with a wide cross-section of local people and gave me an insight into the way they lived. Rich or poor, and we delivered to both, the majority of them were open and friendly and welcomed us as friends and in many instances we were confidants of both their problems and their good fortune. Ours was a family business, carried out by family members, and I was welcomed and accepted by our customers as if I was one of the family. I only once did the Christmas Day round and it was very slow because everyone came to the door to greet us and in many cases to offer us a festive drink.

Travelling the same route at more or less the same time we met and greeted the same people each day. I met my first real girlfriend on the round. She was lodging with one of our customers quite close to my home when I returned to the round at the beginning of my first summer vacation from university. Having seen this extremely attractive girl leaving her lodgings early one morning I contrived to reach there each morning at the same time. Eventually, I plucked up the courage to talk to her and pass the time of day, and in due course to ask her out.

Jillene was a New Zealander on the last leg of a working trip around the world. When I met her she was working as a chambermaid in a large local hotel. When the vacation ended and I returned to continue my studies at Reading, she moved to London where she worked in a Kensington store and shared a flat with another girl from New Zealand. Jill's mother was also in the UK, and she arranged for Jill and I to attend a performance of *Swan Lake* at Covent Garden, the first big professional show I ever attended. What a wonderful choice and what a splendid venue for my first professional show. Then the time came for them to move on and they both returned to New Zealand.

Although the introduction of the milk bottle for our round changed it irrevocably and made it impersonal, I continued to enjoy the work. It was still a valuable experience and the first time that I was entrusted with a real responsibility.

Chapter 13

The Farmsteads

THE buildings that made up both farmsteads were organised around two square yards that adjoined, forming a shape like a figure eight. Here the similarity ended because my home farmstead was larger with more buildings, four of which were very old, much older than any at my second farm.

The narrow entrance to my home farm opened onto a right-angled bend in the road and was flanked by tall, substantial stone gate posts. At one time there had been impressively high gates or, more probably, doors hung there. A large barn with cob walls and a thatched roof flanked the road at each side of the entrance. Wide and tall double doors in the centre of the long side of each barn opened onto the road so that a cart loaded with corn could be driven in, and, in one barn, driven out through a similar pair of doors on the other side into a field. Both had areas of wooden flooring the width of the doorways and stretching the full depth of the barn. One was used as a shearing floor and the other had been a threshing floor when corn was threshed by hand using a flail. When we were clearing out this barn we found a flail. Using it must have been a slow, tedious and tiring task. It was said that an experienced flailman could thresh two sacks of grain a day, whereas a steam-driven threshing machine could thresh 70 bags, seven tons, in the same time! The first successful attempt to make a machine that could thresh corn was made by Andrew Meikle in the 1780s. Farm workers who feared for their jobs and their livelihoods resisted its introduction, and this led to riots in which threshing machines were damaged or set on fire.

At some time a barn threshing machine had been installed in the barn with the threshing floor. It was built around two walls and differed from the travelling thresher that came to the farm. Instead of the crop progressing from the top down, the sheaves were fed in at one end and progressed along the machine. No one seemed to remember when it was last used, but judging by the thick layer of dust that covered it, not for many years.

The barn thresher had been driven entirely by horse power. The horses walked round and round under cover outside the barn. They were harnessed to arms radiating from a vertical shaft which the horses turned as they walked round and round. The central shaft drove the thresher by means of bevel gears and shafting that passed through a hole in the wall of the barn. The radial horse poles had been sawn off to make more room, but the central column and all the gearing were still in place and intact. I cannot recall how many poles there had been but I suspect that at least four horses would have been needed. The remains of wheelhouses, or roundhouses as they are sometimes called, in which horses walked round and round driving machines by means of gearing similar to ours are to be found all over

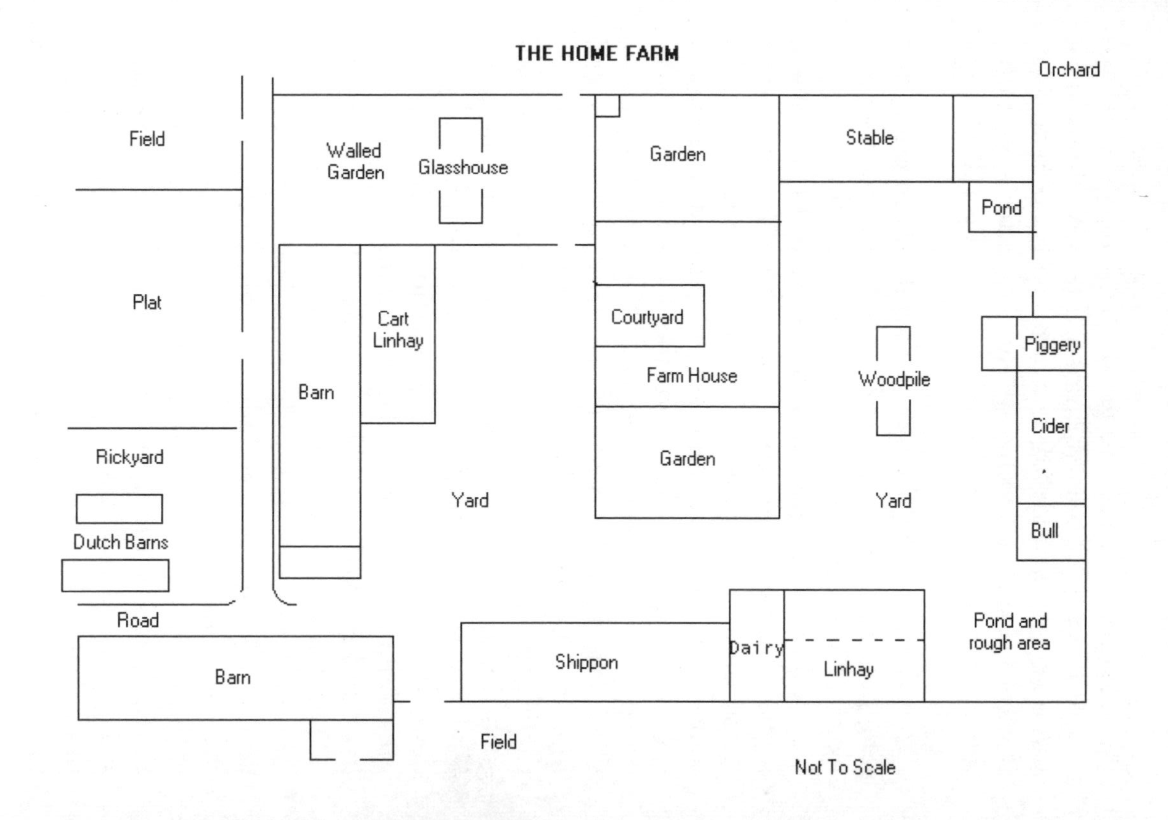
THE HOME FARM
Orchard
Field
Walled
Garden
Glasshouse
Garden
Stable
Pond
Plat
Cart
Linhay
Courtyard
Piggery
Farm House
Woodpile
Barn
Cider
Rickyard
Garden
Yard
Yard
Bull
Dutch Barns
Road
Pond and
rough area
Dairy
Shippon
Linhay
Barn
Field
Not To Scale

the country. There is one in the Yorkshire village where I live. But ours was not a true roundhouse as it was open on two sides and covered with a lean-to roof. It was large enough to house our three carts, two butts, the tractor and other bits and pieces. We called it the cart linhay.

I was present when we removed the thresher and scrapped it to make more use of the space, but the remnants of the driving gear survived until the farmyard was bulldozed to make way for housing. In retrospect, it was a shame that this machinery, together with other items of historical interest, such as a hand-operated winnowing machine, the cider press and vats, a range of field machinery and sundry other ancient items were not preserved. Indeed, the whole farmstead including the house would have made an ideal site for a rural museum. Unfortunately, at the time it was demolished there was little apparent interest in anything old and such things were removed to make way for the new. There was no thought that our history was being destroyed and that within a generation a whole way of life would be forgotten and details of how our ancestors lived and worked would be lost forever.

The farmhouse was also built of cob and thatched. Although not unique to North Devon, cob was very commonly used as a building material in the area, and the large number of cob houses and barns still in existence is testimony to its durability. Originally, all the buildings would have been thatched but many have since been re-roofed with pan tiles. An advantage of all three materials is that they could be obtained locally. Cob was made with clay or loam mixed with water, a little lime, chopped straw, dung and small stones. The mixture was puddled by treading and then laid down in layers 2–3 ft (0.75–1 m) thick. When each layer had dried and set a further layer was put on top, and this procedure continued until the wall reached the desired height. Any protruding lumps were pared away to leave a reasonably flat surface, although they were still uneven and wall papering a bare internal wall was not an option. Provided that the top and base are kept dry, cob walls will stand for hundreds of years, as proven by the farmhouse that showed not a crack anywhere. All the exterior walls were lime-washed white.

The walls were often 3 ft (1 m) or more thick and together with a thatched roof provided a well-insulated building, warm in winter and cool in the summer. Even the internal walls in our barns were as thick. When we removed an internal wall in the cider barn it was wide enough for us to stand on top of it. We had to use pick axes to demolish it, but it was so solid and hard that we could only dislodge tiny fragments with each stroke, and it took a long time and much effort to remove the wall.

The history of the house stretched back many hundreds of years, and there are records to suggest that it was a nunnery before it became a manor house. It consisted of three separate but interconnected wings ranged around the three sides of a small tiled courtyard that contained a well. A high cob wall topped with tiles formed the fourth side of the courtyard. George and his wife lived in the rear part, his father and mother occupied the front part, and the centre part was unoccupied, although one of the upstairs rooms was used as the apple store. All the doors were of simple planked construction with latches rather than handles. All the doorways had lintels much too low to allow me to pass through without bending, and many were the bumps on the head that I suffered as a consequence of forgetting this.

The principal part at the front had a long stone-floored kitchen-cum-living room which was furnished with the long refectory table and bench seats so useful at threshing time. The hearth area was vast with a number of built-in ovens, and the chimney was so wide that the sky was clearly visible when looking up through it. Cooking was done on a range although this was replaced with an Aga while I was there.

There was always a large pan of milk simmering on the range or later the Aga, and when the cream rose to the surface it was skimmed off with a perforated metal saucer. This was the famous Devon clotted cream which was always present on the table at lunch and at teatime at every farm that I visited, and was used in preference to butter and custard. The skimmed milk was sold on the milk round before the war as a cheap alternative to whole milk. Pigs were kept to be slaughtered for home consumption and the skimmed milk and boiled swill formed a major part of their diet. It was much richer than the skimmed milk sold today in the supermarket because the centrifugal milk separator removes much more of the cream than the scalding process. Before the war the farm had sent clotted cream in special tins all around the country by post. There were tins still in store but they were destined not to be used as the home farm didn't re-enter the market after the end of milk rationing. The market never picked up again to the same extent.

The lower yard was separated from the upper yard by the house and the high wall at the side of the front garden. This contained an area of lawn and rose shrubbery; it was fenced off from the top yard by a hedge and low wall. The top yard opened off the road and had the barn and cart shed at the opposite side from the garden, and the dairy and shippon along the third side. The fourth side was yet another high cob wall topped with tiles through which a door led to the vegetable garden and the two outside privies. There were no toilets in the house. The external doors and a few of the internal ones were held shut by simple wooden latches. To open these from the far side you slipped a finger through a hole drilled through the door to lift the latch, or 'haps' as we called it.

The lower yard was sloping, and the central area was always piled with timber that was saved when hedging and which would eventually be sawn up. At one end of the yard there was a linhay, a partially covered open yard for over-wintering bullocks and young heifers. At the other was the brick-built stable with a tallet, or loft, over it. The stable had stalls for four horses and a loose box at each end. Another cob barn made up the fourth side of the yard with a pigsty and a bull house attached. This was the barn in which cider had been made at one time, and which still contained the press and vats when I started. The orchard lay behind it and extended around behind the stables and the house, and out to the road. All the buildings that were not thatched were roofed with pan tiles.

Across the road from the farm entrance was the rickyard. The ricks were built on either side of a stony track. Loose hay, wads of straw, and bales after a baler had been purchased, were stored in two pole barns. These had high single-slope roofs supported on thick poles like telegraph poles and covered with corrugated iron. Three sides were also sheeted in corrugated iron from the roof down to a height of about 6 ft (2 m). It was much easier to stack loose material in these than to make ricks, and stacking bales in them avoided the need for covers as bale stacks cannot be thatched like a rick. However, working in the barns in high summer up under the iron roof was unbearably hot and uncomfortable.

Although the buildings at the second farm were also grouped around two yards, the layout was totally different. The farmstead lay in a dip, and so the entry from the road and the track which led to the fields behind were both steep. The narrow road which served the farm was above the level of the roofs of the buildings on that side of the yard, and the steep exit from the yard split into two forming a Y that made it possible for vehicles leaving to go to the left or right. The farmhouse and a semi-detached cottage and their garden walls extended along one side of both yards. The L-shaped shippon and dairy formed both the division between the yards and the far side of the front one. These two buildings had been recently constructed of concrete blocks with a corrugated asbestos roof. A low, ramshackle storage building filled the fourth side of the front yard against the road. A stable and cattle building formed the fourth side of the rear yard which sloped steeply upwards. There was no building on the fourth side of the yard, and a wall and hedge formed the boundary. The rickyard in the field high up behind the farm contained a single pole barn.

When I was there the farm had no electricity and no mains water supply. A hydraulic ram in a stream some distance from the farmstead utilised the water flow to force water up into a storage tank above the buildings on the far side of the road. From the tank the water flowed by gravity to the taps in the house and buildings. Although this seemed a rather precarious supply for a farm that required a lot of water for the animals and for cleaning in addition to domestic needs, the supply never failed. Once or twice the ram stopped working because of debris blocking it, but this was easily cleared.

Although there was no mains electricity, electric lighting was provided by a generator that was belt driven from the same engine which powered the milking machine. The electricity generated was stored in a set of batteries next to the engine. Although this arrangement provided us with lighting, the supply voltage was low and DC so there was no possibility of having power sockets. Charging the batteries during the two daily milkings was usually sufficient to keep them running. However, in the depth of winter when the lights were on for long periods, there were times when the engine had to be run for longer periods with the belt driving the milking machine removed.

Neither farm had mains drainage. In each case the drains from the privies led to septic tanks, and the top water and drainage from stock buildings was piped straight to ditches.

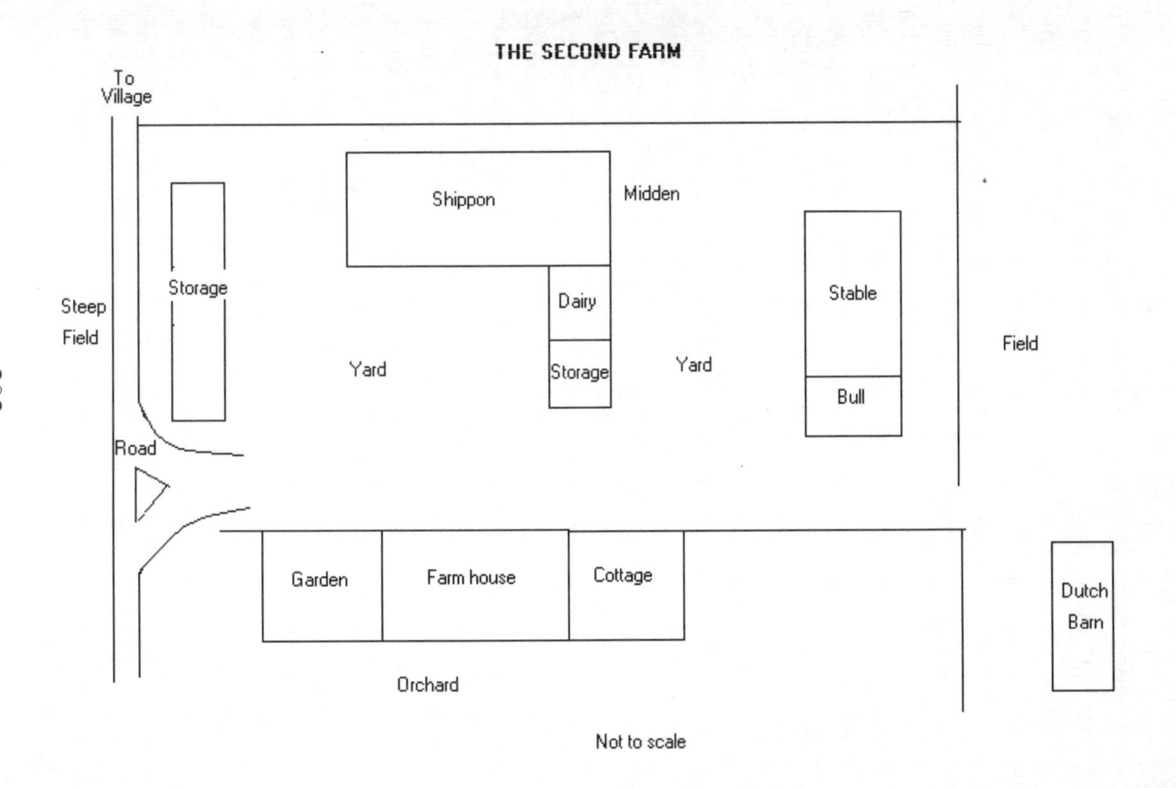
THE SECOND FARM
To Village
Shippon
Midden
Storage
Steep Field
Dairy
Stable
Field
Yard
Storage
Yard
Bull
Road
Garden
Farm house
Cottage
Dutch Barn
Orchard
Not to scale

There were two farm cottages about half a mile apart on the boundary of the land occupied by my home farm. These were 'tied' cottages let to workers on the farm at a low rent for the duration of their employment. When they resigned or if they retired they were obliged to vacate their homes.

Both were built of cob with thatched roofs. One is still in existence, although it has been modernised and considerably enlarged. In their original state the front door led into the stone-flagged main room that, with a lean-to kitchen at the rear, formed the downstairs accommodation. A narrow staircase from a doorway next to the fireplace twisted steeply to the first floor. The landing formed one sleeping area separated from the only bedroom by a flimsy partition wall. There was no bathroom and the only water supply was a cold tap in the kitchen. The privy was down the garden and there was no electricity. Each garden was quite large and was given over to vegetable production. One even had a pigsty.

My second farm had no accommodation for workers, and this may have been why my employer took on a young man with limited experience, and why he applied successfully for my deferment from National Service.

The farms were small in area by today's standards but both would have been described as medium sized for the area in those days. The average size of the fields at my home farm was between four and six acres. There were three fields that I can clearly recall were only two acres in extent and there were one or two plats in odd corners that were even smaller. The exception not included in the average was one huge 30-acre field of very rough grazing. There was a dilapidated two-storey building in the middle of this field with a tree-lined track leading to it. At some time it had been a tallow factory, probably sited there far away from other buildings because of the risk of fire. For some reason that I was never able to discover this field was known as Promised Land.

There were several acres of deciduous woodland more or less at the centre of the farm. It was a source of fence posts, fire wood and thatching spars. A narrow unfenced track through it, gated at each end, was the shortest route to a number of fields. The track was too wet and muddy to be used by carts for much of the year, but at harvest time it was often dry enough for us to pass through. It became overgrown each spring but the ruts made by journeys in previous years remained clearly visible. The narrow track wound sinuously through the wood and so the path was only visible for a few yards in front as you made your way along it. On a sunny day the branches made shadowy patterns on the ground, the birds sang and the only sounds were the jingle of harness and the creak of the wheels. It created a wonderful sense of peace and feelings of freedom and contentment.

Almost all of the fields were rectangular in shape and had quite straight boundary hedges. This would suggest that it was part of an organised restructuring of the landscape which might be the case because I understand that the land once formed part of an extensive manor. It certainly was not part of a communal farming system. Most of the fields were flat but about four sloped very gently.

My second farm was quite different. The fields varied more widely in size and were larger on average. Nor were they as regular in shape, and the hedges wandered from the straight. The farmland lay on one slope of a valley and stretched down to the stream which formed the boundary of the farm and from which the farm water supply was drawn. Most of the fields were steep, particularly at the top of the farm where the land degenerated into heather and gorse-clad moor beyond the boundary. There were a few reasonably flat fields close to the farm on a narrow terrace along the valley side. The valley further up towards a neighbour's farm deepened and so the terrain and our fields at that end of the farm also sloped downwards. Also further up the valley on the other side was a working quarry. The end of the valley beyond our neighbours was characterised by several typical North Devon dumpling-shaped hills rising sharply with steep sides. One narrow road wound its way up and down along the full length of the valley from the village serving both farms before meandering on out of the valley at the other end.

As many of the fields sloped so steeply it was dangerous land for tractor operation. I didn't realise just how dangerous until a man on a neighbouring farm that was no steeper than ours rolled a Field Marshall and was killed. Neither did I realise until at university how potentially dangerous some of our tractor-driving practices were.

One of these was to sit on the wing on the high side when driving across the slope in a field. I regularly did this when mowing and cutting weed, and if I looked down I could see the tyre on the wheel beneath bulging visibly round the side of the wheel rim. From this position I could still steer the tractor, and I felt safer as I had shifted the centre of gravity in the right direction and I could bale out more easily if the tractor did start to go over. The improvement in the centre of gravity was probably irrelevant because although there was a theoretical improvement I doubt if my small weight in comparison with that of the tractor would have made the slightest difference.

One arable field was steep enough to make me change gear on the Fordson Major when pulling harrows up the slope. Normally, harrowing was done in second, but in this field I had to change into first part way up. As soon as I de-clutched the draught of the harrows stopped the tractor, and as they were pulled by a chain there was a tendency for the tractor to roll back. To avoid this I would de-clutch, slam the gear lever across into first and let out the clutch sharply. The sudden jerk as the chain snapped tight and took up the pull of the harrows caused the front of the tractor to rear and the front wheels to leave the ground. Again, it was not until I was at university that I learnt that in those circumstances a tractor could rear so rapidly that it would turn over onto its back before the driver could react and push down the clutch to prevent it, and that fatalities had occurred. Pulling from a high point on the tractor had the same disastrous effect and although I never attempted this I witnessed a tractor that reared and turned over backwards in Northumberland when the driver attempted to pull out a tree stump with a chain attached to the top link. Luckily, this incident had no fatal consequence.

Fortunately, I never had an accident while I was working at my second farm but my employer did. He lost control when towing a trailer down one of the steep grass fields. He had the presence of mind to steer straight down the slope. Had he not done so, but attempted to turn across the slope both the tractor and trailer would probably have rolled over and he would have been lucky to escape serious injury or worse. Braking on wet, slippery grass was usually ineffective as the weight of the trailer pushed the tractor on down the slope and might result in it slewing round. As it was he shot through a sparse hedge at the bottom of the field and down a steep bank into the stream that provided our water supply. No harm was done, although we had difficulty recovering it. I thought that it was a huge joke, but my boss was not amused.

On the opposite side of the road that led past our farm the land rose very steeply. We had one narrow field that stretched along the road in which we pastured our horses and some of the young heifers. Beyond the hedge at the top of the field which was our boundary the land degenerated into moor, with clumps of gorse and brambles. Parts of the hill grew heather, and areas were burnt in rotation each year to improve the grazing for sheep. Even fetching the horses from this field was a struggle, and we used to joke that the stock that grazed there grew legs longer on one side than the other to allow them to stand upright.

My employer decided to re-seed our field bordering the moor to improve the grass. It was so steep that we had to reverse the tractor up to the top of the field and then plough down. Fortunately, it was flat enough at the bottom to turn, but not at the top. Subsequent cultivations were done with horses. Even this was a problem as the harrows sidled sideways and the horses kept slipping. We gave up after cultivating one end and fertilising and re-seeding it by hand. To the best of my knowledge no attempt has been made to plough that field since.

My employer had several fields near Croyde where he had originally farmed. As they were so far away they were cropped while I worked for him. He did most of the work there himself but on one occasion I had to bring back the tractor and plough. He took me there by car and I had to find my way back along the main road through Braunton, right down Boutport Street en route through Barnstaple, and back to the farm. The Fordson Major had its reversible mounted plough fitted. The plough projected a long way out from the back of the tractor. The effort of balancing it took weight off the front wheels so that the steering was light but decidedly skittish. Coupled with the fact that turning to the left resulted in the plough swinging in the opposite direction, plus my limited experience of driving with mounted equipment, several parked cars almost had their back or front remodelled as I weaved my way through the busy, narrow streets. More by good luck than judgement I negotiated all the hazards and arrived safely back at the farm.

I had been given a set of driving lessons as an 18th birthday gift just before I went to work full-time at my second farm in July. My employer gave me some time off to take them as it would allow me to drive the tractors on the public roads. My parents had no car and so all my instruction and practice was provided by the driving school in the High Street. The car used was a pre-war Austin two-seater with a dickey seat. I passed the test just before Christmas.

After I left home for good there were major changes at the home farm. The town had advanced to the edge of the farm and the farmstead and land around it were sold for building. The old farmstead was demolished and a new set of buildings erected on the field known as Promised Land. The new set of buildings accommodated a modern well-organised dairy unit. George sold the farm several years later; and soon after, the new farmstead and all the farm land were swallowed up by an industrial estate and housing. Many times since then George and I, accompanied by our wives, have walked the new roads and streets trying to recall where field boundaries were and swapping memories of incidents that occurred while we worked in them.

My second farm has changed little although the road to it seems narrower than it did, probably because grass has grown down the centre and it now looks to be little more than a track. Of course there is no road man now to care for it and the limited funds available for road repairs are directed to major routes and little used lanes are neglected.

Chapter 14

Leisure

DURING the war and the years following it while I was still at home our leisure time was spent very differently to how it is today. For one thing there was less of it. Most people, and certainly those who were considered to be working class, worked longer than they do now and for many the work was more arduous. Not only were their working days longer but the working week was usually five and a half days. Two week's holiday plus five Bank Holidays was the maximum annual holiday that could be expected, and many received less. When I left school and started at my second farm, I began work at 7.30 am and finished at 5.30 pm, 12.30 pm on Saturdays. The journey at each end of the day took a further half-hour.

Even my free time while still at grammar school was limited. We were given homework in two subjects each day that sometimes took a couple of hours to complete to the standard I had set myself. At weekends we were given three subjects, and I usually tried to finish them on Friday evening to leave Saturday clear to spend at the farm.

The degree of manual labour and the lack of mechanical aids to reduce handwork in farming have been described but the same reliance on physical labour extended to most work situations. There were no JCBs, powered lifting and loading machinery, and few hand power tools available. If a water main had to be installed the tarmac was broken with noisy and heavy vibrating pneumatic drills, and the trench was dug by hand with pick and shovel. When house building the mortar was usually mixed by hand, every brick, timber and tile had to be carried up ladders to the upper floors, and all the timber work and fitting was done with hand tools, saws, drills and planes. There were no power tools. At the end of a long, physically exhausting day most men looked forward to a restful evening at home or at the local pub with their mates.

The limited leisure time was matched by the limited ways of spending it, particularly for those living in the country. There were two cinemas in the town and a small theatre that had a rep at one time and was also used by amateur theatrical groups. I occasionally visited shows there. It was at this theatre that I was introduced to light opera and to Gilbert and Sullivan in particular. Many years later I was to take part in amateur operatics as a member of the chorus and sang both Gilbert and Sullivan and other light opera. Discos were a long time in the future, of course. There were probably dance halls in the town, although my Methodist upbringing wouldn't have favoured them if there were. I didn't learn modern and old-time dancing until I reached university.

Transport was very limited as well. There was a bus service on the main road from Barnstaple to Bideford, although the nearest bus stop was half-a-mile distant from my home. Getting home after any late function was a problem. We were fortunate that we lived only a couple of miles from the town and were able to walk home if the last bus had gone.

Few people owned private cars, and most of them spent the war up on wooden blocks in the garage to take the weight off the tyres, as petrol was only provided for essential purposes. We had no car, and the only times that I travelled in one until the late 1940s were when my uncle who lived in Launceston paid us a visit. He was a timber valuer. Travelling to assess standing timber was of national importance and so he had an allocation of petrol.

Workers lived close to their employment and walked, cycled or travelled by bus to get there. I cycled almost five miles each way to the farm where I worked and it was all up hill and down. There were few flat stretches on the route and it took me half an hour, pedalling hard for much of the way, to complete the journey.

The coastline and river estuaries were closed and guarded during the war for fear of enemy invasion, and latterly because they were used for D-Day training. So leisure outings were limited to walking or cycling through the country lanes. On Sunday afternoons when the weather permitted we always walked off our lunch in the network of lanes and roads near our house. There was no traffic except the occasional bicycle and from the top of Roundswell we could see Codden Hill in one direction and the estuary of the Taw in another.

There were a few shops half a mile from my home but most of the shopping was done in town. My mother carried everything home on the bus and then walked the last half-mile. However, she didn't have to carry vegetables because in common with most of their neighbours my parents grew enough to provide us with a succession throughout the year. In addition to our large back garden, Dad had an allotment in the grounds of the Methodist Chapel not far from our house.

We kept poultry in a run in the back garden to supplement our food supplies. They supplied us with eggs, and ultimately made the final sacrifice for Sunday lunch. Hens only lay for part of the year and so my mother stored any surplus for the winter in an earthenware jar filled with isinglass. I never remember us being without eggs throughout rationing.

We were able to buy a ration of meal for the hens in return for surrendering our egg ration. The meal also had to be carried home on the bus. The hens' diet was supplemented with kitchen scraps that were boiled in an old saucepan, and grain that we were allowed to glean in the cornfields after the corn stooks had been carted and the field raked and cleared. Mum hatched a sitting of eggs each year under a broody hen, or bought day-old chicks to raise in order to maintain the number in the flock.

After the war Mum decided to give up the hens. When the houses and run had been dismantled Dad planted potatoes in the area they had occupied. The resultant crop had the tallest and most luxurious haulm anyone could recall. Unfortunately, the crop of tubers was no greater than normal. Undoubtedly, the lush haulm was the result of the nitrogen residues built up as a result of years of nitrogen-rich poultry droppings incorporated in the soil.

There was no TV of course but we listened to the radio instead, although there were only two official channels available, Home and Light, plus a pirate station, Radio Luxembourg, which played the popular music of the day, if our aerials could receive it. Our aerial was a long wire that stretched from the eaves to the top of a pole at the far end of the garden. I recall very clearly serials such as *Dick Barton, Special Agent* and *Paul Temple*, and dramas and documentaries like the story of Gilbert and Sullivan.

The 'pop' music, if it could be called that, was very different from what we hear today. Rock and roll had still to burst onto the music scene and we were still in the era of big bands with vocalists that played dance music, swing and traditional jazz. I enjoyed these and the light orchestral music, songs and ballads on the Light Programme. The feature common to all the music of the period was that there were recognisable tunes that were easy to memorise and to sing or whistle. I cannot remember the last time I heard someone whistling in the street but in those years it was commonplace, particularly amongst deliverymen and errand boys on their bicycles. The latter too have disappeared from the streets.

Much of my family's social life was focused on the Methodist Chapel nearby. Dad had a nonconformist background but Mum was brought up in the Church of England. However, the Chapel only a quarter of a mile from home was newly built and had a lively congregation and they became members almost from the time that it opened. By the time I was in my teens Dad was a member of the choir, chief steward and treasurer, posts that he held for many years. Mum was convenor for 'faith' teas and also sang in the choir on special occasions. She was also a member of the women's fellowship and specialised in reading Jan Stewer stories in Devon dialect. This was perhaps rather odd as she was born in Dorset and lived there until she married my father. However, her home was only a couple of miles across the county boundary and AJ Coles, the author of the Jan Stewer stories, lived and taught close by in East Devon, so her accent was probably authentic.

There was an active social club for the young people of the church which met each week, although there were only half a dozen of my age in the area. From time to time social evenings were organised and people of all ages took part in the party games and quizzes that were organised. Faith teas or suppers were part of the occasion, although not too much was left to faith as there was a convenor who made sure that there was plenty of food forthcoming.

We young people were encouraged to take part in services and to join the choir when we reached our mid-teens or in the case of the boys, when our voices broke. I learnt to be able to read and sing the bass line by singing next to my father, although I could read music as I had been taught to play the piano when I was younger. Dad was a member of the choir for a number of years and was a very accurate if not powerful singer, unlike my grandfather who was a fine self-taught musician with a huge bass voice who had taught himself to play both the piano and the organ.

After the service on a Sunday evening, our small group of young people would regularly go to one another's homes where we would often gather around the piano and sing. Two of the group were accomplished pianists. We had a piano at home, too, although I was an indifferent performer. Choir and youth club outings took place annually, and each year we went by coach to places as far away as Torquay. These were hugely popular and very much enjoyed by the people of all ages that went. They were the only opportunities that most of us had to travel about. It must be remembered that in the whole congregation there were only half-a-dozen car owners, almost all of them men who had to have cars to conduct businesses which were of sufficient national importance for them to be issued with petrol ration coupons. Even relatively close places of interest were only accessible to us on outings such as those organised by the Chapel and similar organisations. During the war years, travel was even more limited as all of the coastline and beaches of the West Country, like those of the rest of the country, were closed off and festooned with barbed wire for fear of invasion. Amphibious vehicle training took place in the Taw estuary, and the beaches were used for D-Day training.

One of our family friends was a local preacher and travelled around the Bear Street Methodist circuit preaching at the different chapels. There were two Methodist circuits in North Devon in those days, the second based on the second large chapel in Barnstaple in Boutport Street. Our friend was a market gardener with a stall in the pannier market held in Barnstaple each Tuesday and Friday. He conveyed his produce to the market with a car and trailer and journeyed to the more distant chapels in the circuit, which stretched from Barnstaple to the edge of Exmoor, on a Sunday.

Several of the chapels served tiny communities in remote areas and only had small congregations. When our friend was scheduled to preach in one of these he would often take me with him to 'swell the congregation and help the singing', although most Methodists seemed to be enthusiastic, if not always tuneful singers and enjoyed 'making a joyful noise unto the Lord'. By this time I was in my late teens and was working full-time on my second farm. I looked forward to these outings with him and not least to the journeys.

Riding in a car was still a rare event for me, although I was learning to drive, primarily so that I could drive the farm tractors on the road. His car was a pre-war Morris with a huge square body and masses of room in the back. It was far from quick but I enjoyed the ride, particularly in the dark in the winter as we travelled home along deserted narrow and winding lanes lit only by the tiny pools of light from the six-volt headlamps.

Frequently, because of the difficulties that preachers had in travelling to the more remote chapels, he was booked for both morning and evening services at the same chapel and we would be invited to spend the time between the services with a family from the congregation. These were memorable visits. They always began with an enormous roast lunch with a joint and masses of different vegetables followed by a pie with, of course, clotted cream. This had all been prepared before our hosts went to Chapel and the roast left to cook in the range or Aga. After a pleasant afternoon spent chatting or strolling around the farm we would assemble in the dining room for tea, another formidable meal with ham and salad, several different cakes, scones and jam, trifle and, of course, cream again.

Several of the farms were remote and one near Challacombe on the edge of Exmoor was at the end of a muddy track through several fields with gates which had to be opened and closed as we passed through. That was my job as passenger, and in winter it was usually raining! I remember particularly this farm, another at Goodleigh and a butcher's house in Bratton Fleming. There was a wonderful friendly Christian atmosphere at every house that we visited. We were always made most welcome and all the women seemed to be wonderful cooks despite the limitations caused by rationing.

While working on the farm I had more free time during the evenings than I had at school, particularly in winter. At school, homework had occupied much of the evening.

The working days were quite long, however. I was up in time to demolish a cooked breakfast before I left at 7 am to cycle to the farm. I usually arrived home again at about 6 pm ready for the substantial meal that my mother had cooked for me. So I didn't get out again much before 7 pm, and late nights did not go well with early mornings. In spring and summer the working days were sometimes much longer. If the weather was good and there was field work to be done we might continue until dark.

During university vacations I tended to take all the overtime that I could to supplement the county major scholarship that financed my three-year degree course at Reading. George was keen for me to do so; we worked well together and enjoyed each other's company and we often worked on after the other men had knocked off.

Both while working on the farm and during university vacations I continued to sing in the choir with my father and accompanied him to practice each Thursday evening. We both joined a Free Church choir that had been assembled to sing *Elijah* in the Queens Hall in Barnstaple. The choir soon became the North Devon Choral Society and we went on to perform *The Messiah* in the Queens Hall in Barnstaple and again at Ilfracombe. Dad continued to be a member of the Society for many years after I had left the area for good.

On one other evening each week during winter I used to cycle into town to attend evening classes in agriculture at the local evening institute. This was the first formal instruction in the subject that I had received, although I read the *Farmers Weekly* regularly and had been given one or two rather out-of-date books on farming. The classes were conducted by staff from the Devon County Farm Institute at Bicton in East Devon. Many years later, I derived great pleasure from visiting the college as a national examiner for a course they offered in agricultural engineering. By coincidence, some years later my cousin married a lecturer at the college and he became responsible for the instruction given at the out-centres such as Barnstaple.

One event in the calendar that people of all ages in North Devon looked forward to was Barnstaple Fair. In those days it took place in an area bounded by the town station and the cattle market and the roads in the area were closed. Although the showmen's engines and trucks towing trailers loaded with the rides rolled into town from Sunday onwards, they were not allowed to start up their rides and side-shows until after a formal opening on Wednesday lunchtime and they had to close down again at midnight on Friday.

It was said to be one of the largest fairs in the west of England. To give an idea of its size there were often several sets of dodgems, waltzers and two big wheels, in addition to a host of other rides and rows of side-shows and stalls. The fastest ride of the period was the moon rocket, a set of rocket-shaped carriages which rotated on an inclined plane and it was boasted reached a speed of 90 miles per hour. This now seems a very modest speed but at a time when only railway engines reached that speed on the ground it seemed incredibly fast and the nearest thing to flying. Our evenings were rounded off with fish and chips – another rare treat – and as we walked home up Sticklepath hill we would look back over the river at the masses of coloured lights that twinkled on all the rides and stalls. The big wheels always stood out high above all the other rides. It was a wonderful sight after a period of blackout and austerity and when shops and streets were not lit up as they are today. The fairground was packed with people each evening and everyone we knew seemed to be there. For a few getting home was a problem. We could walk but I know of someone who missed the last train back to South Molton and walked the 12 miles home along the railway track!

The day after the fair closed was Carnival Day and a procession of decorated floats wound its way around the narrow streets of the town. The pavements along its route were all packed with people. The only other processions around the same route that I can recall were more sombre affairs and took place on Remembrance Day each year. My father had served in the First World War and had led an ARP party in the 1939 war and so he always took part in the march.

Although our leisure activities may seem to have been limited in comparison with the wide range of opportunities available today, we were never short of things to do. Despite the hardships of the war and the immediate post-war years we were able to enjoy our youth. I certainly don't regret that I was born in the thirties rather than at the end of the century.

About the Author

Born in N. Devon in 1932, Mike Hawker is descended from many generations of blacksmiths and farm workers who lived in E. Devon.

The author's interest in farming developed over more than a decade of working on farms, first as a helper and part-time employee and then as a full-time worker. He took a degree in agriculture at Reading and an M.Sc. in agricultural engineering at Newcastle. Three years research and a PhD led to lecturing posts, first at Edinburgh and then as head of engineering at Askham Bryan College. Before retiring he became senior manager responsible for the estate and the buildings on the college sites.

Dr Hawker is an Honorary Member of City and Guilds. He served on numerous committees and was syllabus writer and examiner in both horticultural mechanisation and agricultural engineering. He visited Ghana on behalf of City and Guilds and the British Council to develop examination systems in agricultural engineering, and was joint author of a textbook on horticultural machinery that was widely used in colleges for over twenty years. Dr Hawker was also national examiner for several other examining boards and was an active member of the committees of both the Agricultural Training Board and the Road Transport Board.

Although his heart still lies in Devon, Mike continues to live in Yorkshire. He is married, with three daughters and six grandchildren.

Contents

Acknowledgments

The idea for this book and its partner, *Self-esteem for Boys*, came from my editor, Jacqueline Burns. Had she not been such fun to work with, I may not have gone ahead with it. So I'd like to thank her, and Amy Corzine, who took over the reins for Jacqueline's maternity leave.

I am indebted to two teacher friends in particular, Dexter Hutt, Principal of Ninestiles School, and Gary Wilson, Head of English at Newsome High, who both gave precious time to long conversations and script reading, which I needed and really appreciate. I should also like to thank Geoff Evans, of the C'mon Everybody project; Adrienne Katz, of the research organisation Young Voice, which published *The Can-Do Girls*; Jo Adams, for her report *GirlPower*; and Alex Vear, a good, wise friend and mother of two daughters in the thick of growing up, for helping in various ways.

Finally, I'd like to thank my daughter, Georgia, who is just entering adolescence, for her constant support, love and understanding. I wish her well for the adventure ahead.

CHAPTER 1

Understanding Her Challenges and Opportunities

Everything seems to be going right for girls. Their confidence, exam results, and career opportunities are rising. Released from the prison of home and hearth, they have a freedom that was unheard of 50 years ago. Many boys feel that girls have it easy and resent it, for the have-it-all world that beckoned their mothers seems a go-and-get-it-all world to their daughters, and leaves boys behind.

Competent, successful superwomen have bred competent, successful supergirls – or have they? Not very far under the surface of 'girl power' lie new pressures and insecurities that come with our daughters' enhanced potential for 'success'. Some, such as eating disorders, have arrived in the slipstream of that success; others exist where success is absent, for success has certainly not come to all girls, who today manage a triple dose of expectation: to be academically and economically successful; to be emotionally independent; and to achieve the heightened standard of beauty that cosmetics, clothes and even surgery can now provide. They are expected to look good for themselves and their female friends, not necessarily for the opposite sex, who are regarded as bringing more problems than they are worth, hence the new female goal of emotional as well as economic independence. Alongside this emotional aloofness is an equally pervasive pressure for girls to be not only sexually aware but also sexually skilful. The magazines tell them how. It is an emotional minefield in which many vulnerable girls get blown up, as teenage pregnancy figures testify.

There are academic casualties, too. While many girls are working the new educational system to their advantage, some succeed only at considerable cost to their mental health and well-being. Others fail to make it at all. Two to three girls in every 100 will attempt suicide some time in their teenage years. Indeed, females between 15 and 19 are the highest single risk group for attempted suicide.

Some parents add to the pressures their daughters feel. Like them, they fall prey to the look-good, feel-good, get-rich-quick society. Mothers can look more stunning than their ugly duckling daughters for years, and must be very self-assured not to play this game.

The brighter side of the story is that at the beginning of the twenty-first century, females have more opportunities, and are less restricted by their gender than ever before. There are few areas of sport, leisure or work that remain the exclusive preserve of males, in theory at least. They can have all this and also become mothers.

With so much available to girls, parents and teachers have the task of helping them make the most of their abilities. However, to maximise their chances and to protect themselves against the potential problems of their new position in society, girls need plenty of positive self-esteem and large doses of genuine self-assurance.

Girls must fashion a new femininity – one that will encourage integrity, respect for others and themselves, independence and autonomy while acknowledging their biological inclination to nurture. Femininity used to imply making oneself look pretty for men, being self-effacing, submissive, compliant and placid, always thinking of and caring for someone else's needs before addressing one's own. This is now rejected by most young women but, in pushing themselves forward, some have simply aped men.

A respectful regard for themselves as being female is vital if they are to remain psychologically intact in the face of the abuse that so many suffer, and also for their female gender. They should not consider themselves weak or inferior to boys because they may be more emotional or don't always feel in control. Girls must acknowledge their own needs while recognising other people's, see themselves as having worthwhile views, and accept that they deserve care and respect from others.

The amount of self-criticism, low self-esteem and dissatisfaction with their looks and performance that exist unexpressed behind even confident exteriors should be a matter of considerable concern. Girls thrive when people notice when they've done something well, listen to and take them seriously, and acknowledge their rights.

If girls are to become more self-assured, parents and other involved adults cannot start too soon. Building self-knowledge, identity, confidence and self-esteem – the constituents of inner strength – is the way to create a resilient young woman who is able to cope with challenges and an uncertain future with confidence.

CHAPTER 2

Meeting Her Needs

Girls' fundamental needs are the same as boys'. No one has an awareness of their gender until the age of about two, and most of us go on needing the same sort of essential support and care from those close to us throughout our adult lives. In addition to being fed and clothed appropriately and kept clean and healthy, very young girls, like boys, also need emotional sustenance. They need to be loved and cherished, appreciated and valued, noticed, enjoyed and admired. Girls thrive when these needs are met because they feel important and significant. When they feel secure, competent and capable, when they are listened to, especially as a source of authority in relation to themselves, and when they are able to develop their talents, they grow up feeling strong inside and able to walk tall. By contrast, if a girl's basic needs are not met, she will feel neglected, separate, unworthy of attention and full of shame.

Growing girls have special needs on top of these basic ones that relate to their future role in society. Parents and teachers should attend to these as well. Girls now require the confidence and flexibility to balance the demands of work and family and, at the same time, to manage the uncertainty implicit in likely domestic and job changes. They should grow up able to love and trust without making themselves vulnerable to exploitation and abuse. Girls should be helped to become strong enough inside to defend themselves against any assault or intrusion on their physical and mental integrity, at home or at work.

During childhood, girls typically develop closer relationships with their mothers. When mothers are reluctant to let go, the possessiveness and competition that arise may make separation that much harder and more painful. It is not uncommon for intimacy to be associated with exploitation and emotional aggression. Intimacy should not be allowed to develop into dependency or domination.

Meeting our daughters' need for intimacy, succour, honour and autonomy will take time and considerable effort. But if we don't put in that effort, we are likely to have to work much harder later on picking up the pieces of their consequent emotional distress.

1 Help her to believe in herself

As I left the house for my first date, my father told me that I looked wonderful and the boy was very lucky to be going out with someone so special. If he didn't appreciate that, he was the loser. He made me feel the tops and gave me the strength to cope with anything. – Jo Brand

If a girl has good self-esteem, she will automatically believe in herself as a capable and lovable person. She will have a clear and positive sense of who she is and will see herself in a favourable light. Girls with self-belief will be optimistic about what they can manage and achieve; they will be able to raise their sights, stand up for themselves and explore their potential. People who lack self-belief are filled with a generalised self-doubt which predisposes them to feel guilty, shameful and inadequate.

No girl will believe in herself if someone has not first demonstrated belief in her competence and capability and considered her worthy of love, support and attention.

Parents

- show faith and trust in your daughter – in her ability to decide certain things, to succeed at tasks, to manage her own personal care and be responsible when she's old enough
- show her that you love and enjoy her in ways that will convince her
- beware of thinking: 'I didn't get or need support, so she'll be the same' ; she is different; also, people who don't get often turn this into 'don't need' to cover up their own disappointment
- love can be shown physically in ways other than kisses and hugs – by, for example, sitting close to look at books, magazines or television, or on her bed at bedtime

Teachers

- teachers will not feel parental love for the girls in their class; however, they can make it clear that they enjoy, approve of and accept their students for who they are
- identify two or three girls in your class who show signs of low self-esteem; try to talk regularly to each one and comment favourably on her various attributes and abilities
- giving special tasks to these girls can make them feel significant, noticed, reliable and trustworthy

2 Show that you understand her

My six-year-old came home from school one day particularly fractious and tired. It turned out she'd had a difficult day squabbling with her two best friends. She felt really low, but when I told her about 'Two's company, three's a crowd' and explained the difficulties and dynamics of threesomes, even for adults, she visibly brightened.

All children find being misunderstood hugely frustrating. At first, they'll just be irritated, but when the mistake persists, they will begin to doubt themselves and their sense of reality. If your daughter's wishes, thoughts or experiences are continuously ignored or misinterpreted, it won't be surprising if she becomes resentful and angry.

You can show that you understand her by anticipating her needs and expressing her likely thoughts – though carefully – using phrases such as: 'I guess you're feeling a bit left out, am I right?' Posing the question at the end gives her room to disagree and stops you coming across as infuriatingly all-knowing – and possibly wrong.

Parents

- accept the way your daughter sees the world – she does not have to agree with you, nor you with her
- value her uniqueness; tell her what it is about her you admire and treasure
- look beyond her behaviour at possible causes and feelings
- repeat what she says to you, so that you can make sure you've understood correctly: 'So you want me to stay in tonight because you're fed up with me working late so much this week, right?'
- remember her likes and dislikes
- state what she's likely to feel: 'You won't want to hear this, but I can't afford a new jacket for you this month'

Teachers

- make a conscious effort to see patterns in a girl's work that might show you what makes her 'tick'
- encourage class work that draws on students' likes and dislikes, and try to remember a few of them
- for girls who seem particularly trying, list four reasons why this might be so, excluding 'difficult personality'
- use 'reflective listening' phrases: 'What I hear you saying is that you did not feel you knew enough to start this homework. Let's start from what you're sure you do know'

3 Approve of who she is, even if you dislike what she does

Every girl needs to be accepted and approved of for who she is, not just because she has been 'good', 'helpful' or 'successful' and lived up to your ideal of who she should be. If she constantly fills a mould set by you, she'll quickly lose her identity and find it hard to feel happy about who she is.

Young children are always getting into scrapes because they are learning about rules, how things work and how to manage themselves. Clumsy reprimands convey disapproval and can do great damage. If you want to comment on something your daughter has done, be clear that it's her actions you disapprove of, not her. This will leave her self-worth intact while she learns to manage her behaviour and appreciate what flows from it.

Don't lose faith in her just because you are unhappy about a particular act or attitude, or leave her feeling devastated by your criticism.

Parents

- think about your daughter's good points before you criticise a particular behaviour, to help you think positively and make your comment specific

- avoid using the words 'good' or 'bad' about her behaviour, because she'll take them as reflections of herself; instead, talk about what it is she does that you like or dislike

- limit your disapproval to the moment by saying, 'Right now, I find you...'

- striking her with your hand or an object will encourage her to feel that you dislike her and she may then decide she is not good enough to be liked

Teachers

- identify something you like about each student, then it is easier to state honestly that it's her behaviour that's the problem, not the girl herself

- describe in detail the behaviour that is outside the rules; and avoid 'You' statements; saying: 'I'm finding the way you are tapping your ruler irritating' is less personally offensive and provocative than 'You are being really irritating!'

4 Give plenty of praise

Children love to be noticed and give pleasure. It is lovely to see our daughters beam with pride when they have done something well and they know we have noticed. This is the essence of constructive praise.

Girls enjoy praise because they like to know that someone appreciates the effort they have made when they have tried. But praise also helps to develop self-discipline. Through praise and encouragement, girls receive clear, positive messages about how they should lead their lives – what it is they should do, instead of hearing what it is they should not be doing.

Many people find it hard to be generous with praise. Being critical makes them feel in charge and all-knowing. Praise, on the other hand, can make them believe they've lost that powerful edge. Some don't know what to praise or what words to use. Others believe that praise will make a girl big-headed, or lazy and over-satisfied with work that isn't perfect. But being noticed and appreciated usually makes girls try harder, and it shows them how to give positive feedback to others.

Parents

- say things like :'That's great!', 'Brilliant!', 'Well done!' and 'Thanks, that was really helpful'
- find something to notice and praise at least once every day
- girls can be praised for their thinking skills (their choices, ideas, solving problems), social skills (helpfulness, understanding, sharing and flexibility), physical skills (running, tree-climbing, making things, sport) as well as for pleasing reports and marks
- be specific: praise what she has done, don't effuse in general about how wonderful and clever she is
- able girls also need their effort recognised by their parents, even ifthey usually do well

Teachers

- encourage girls to evaluate and praise their each other's work, so that praise doesn't always come from someone in authority
- help students to feel satisfied: 'I expect you felt really pleased with this when you finished it'
- find something to praise every day, and include humour, sociability and creativity
- if a girl rejects all praise, showering her with it won't work; select one thing you find truly pleasing, and repeat it three times every day for three weeks, so that she begins to believe and trust that it is true
- take care: public recognition of success may lead to perfectionism and praise-dependency

5 Spend time with her

I know my dad loves me, but I hardly know him. I know he works hard to support us, but we hardly ever talk. It makes me feel as if I'm incomplete.

Research shows that children like to have their parents around, even if they're not actively doing anything with them. Girls like to see fathers as much as mothers where possible, and even teenagers are reported to want to see more of their parents, even if it is a case of being 'seen but not heard'!

British men and women work longer hours than in most other countries in Europe. This means they spend less time at home with their families. Girls cannot feel loved and lovable, believed in and believable, respected and respectable, if the people on whom they depend seem not to care. Only if the important adults in a girl's life give her time and attention can she feel validated and develop any kind of self-worth. Meeting your daughter's need for stress-free time with you, when you give her your undivided attention, will help her feel confident and significant.

Parents

- presents are no substitute for presence: don't try to buy your way out of being unavailable
- spend time finding out what your daughter thinks, and talking about what she has been doing
- play with her, watch her doing or join in her favourite activity, or say: 'I'd love you to talk to me while I wash the car/peel the vegetables', etc.
- keep every promise to visit, and stay in regular touch
- quiet time together can be as valuable as action-packed time
- try to do any office work you need to do at home in family space so that you're not cut off
- put the answerphone on

Teachers

- don't let bad behaviour, or learned helplessness (to which girls are particularly prone), be the only way to get your attention
- if a student wants a conversation at an inconvenient moment, suggest another time when you can listen to her properly
- each week, identify the retiring girls and organise with colleagues to make a special effort to speak to and engage with each one every day

6 Communicate with touch and words

Touch may often convey what you want to say better than words. It is far less open to misinterpretation, and need only take a second. The positive touch is, for example, a full embrace or an arm round your daughter's shoulders; it can ask for nothing in return or seek a simple sign from her that she feels the same way. It can show to her and others that she belongs to you. It can heal an argument and say you're sorry. It can console her after a disappointment, demonstrate your pride or be a show of equality and partnership.

So touch can reassure as well as relax. But it can also hurt. Hitting even a young girl will usually hurt her deeply, and simply pushing your daughter away when you're angry can be a signal interpreted by her as deep rejection. A girl who is never touched can feel ignored and ashamed of herself, and may become easy prey to inappropriate attentions from others.

Parents

- little strokes of your daughter's forehead, head or hands at bedtime or while watching TV – or just sitting close – can be a way to get the habit of touch back into your family if it has disappeared
- experiment with using touch as an alternative to words
- some children don't like too many cuddles; don't force it, just find other ways to get close, to experience togetherness and show your love

Teachers

- child protection issues make many teachers reluctant to touch children; in any case, as girls grow older, it becomes increasingly difficult and inappropriate to do so; just standing close to a student as you look over her work can show you accept her and feel no discomfort in her presence
- some teachers greet their class of young children individually as they enter the classroom, inviting each pupil to choose how to say hello each time – with a smile, a handshake, or nothing at all if that's how they feel that day

7 Respect her right to know

Most children thrive when they feel secure and can predict what is going to happen to them. The unexpected can be very unsettling. Sometimes things happen out of the blue and any adults involved can be equally surprised. But more often the adults know in advance and simply fail to keep a child properly informed.

Children need to be able to make sense of their world. If they can't, they live in social and emotional chaos. They make sense of life both through the patterns that emerge when life is ordered and each day has a predictable shape to it and through being given explanations when changes occur. Young children's brains develop by constructing meaningful patterns, so every child needs to make sense of knowledge and events before she can learn.

When you explain things to your daughter, you show that you respect her right to know, empathise with her need to make sense of her world, respect her ability to understand, and trust her with the information.

Parents

- try to tell your daughter about things before they happen, and as they happen, and explain afterwards why something happened
- tell her about your own feelings, and discuss hers
- she can be told about variations in routines, when partners and relationships change, and have any absences explained
- inform her about decisions taken, and the reasons for them
- give her the facts, answering her questions about such things as death and divorce, honestly, but in terms which she can comfortably comprehend

Teachers

- give girls good warning of any changes to classroom routines
- if you know that you're going to be away, give them notice, and let them know who'll be taking your place
- explain why any punishment or 'consequence' is being imposed
- explain why a piece of work is either good or falls short of the required standard
- keep girls informed about the time it will take to mark important tests or projects, and explain any delay in returning work

8 Shore up her sense of self-worth, constantly

Girls are like sand-castles. One minute they stand perfect, proud, intact, seemingly impregnable against the enemy; the next, the tidal waters of self-doubt are lapping away at their foundations, causing a progressive collapse. Like Canute, we can't hold back the tide, but we can be on hand to dig furiously to reinforce our daughter's self-worth, so that when the tide recedes, there is something left. If we have enough knowledge and foresight, we can help her to establish herself above the tide-line, but it's rare that safe territory can be spotted in advance. There are no rock-like certainties in the world in which our girls live today.

The waves that erode a girl's foundations are fads and crazes, body image and fashion, pressure to be an academic or social success, and sexual activity. Just when she needs more than ever to be sure of herself, given the demands and uncertainties within her life, these forces try to flatten her into a relentless uniformity. The more she loses her true self, the more she has to rely on image and peer approval to feel comfortable.

Parents

- try not to let your daughter become obsessed by collectables and fads; help her to remain her own person
- never refer, either approvingly or disapprovingly, to her weight or body shape, unless it is to counter her own negative comments
- always balance any comments on her looks or appearance by saying that it's what's inside that counts
- if she says: 'I won't be popular unless I...' (follow some trend or other), it's a sign that she's feeling insecure; giving in will make her more reliant on conformity; suggest reasons she can give to friends, so she can resist them with more confidence

Teachers

- be vigilant to spot any signs of low self-worth in students; step in with positive comments and approval if you hear them say anything self-deprecating
- encourage class discussions about any craze as it emerges, and stress constantly the importance of making up their own mind about what they like, want and need
- if a girl says she is no good, is stupid or knows nothing, challenge this, not with an outright denial, but by saying how you find her work, e.g.: 'I find your work full of good ideas', or, 'I would not have been able to give that essay a B if you had known nothing'

9 Be her last refuge

When your daughter has been going through a tough time, when she has had enough and has no energy left to keep up a front, she'll need somewhere to hide, somewhere, and someone, to be her last refuge. This is the place where she can be herself, where – for a short time anyway – no one will be holding her accountable and she is accepted, unconditionally. Here she can truly relax, safe in the knowledge that someone is there for her, someone who will share her burdens for a while and give her some relief.

Home, for children, is the obvious place, and parents are the obvious people, because they matter, though there may be times and situations when parents feel they don't have the emotional reserves to give the consolation required. You might, though, consider whether you daughter really needs anything except to be close to you, without saying or doing anything else. It need not take much to help her renew her faith in herself.

Parents

- if your daughter asks for forgiveness, accept her olive branch and try to put any difficult incident behind you
- giving her refuge does not mean you have to ignore forever behaviour that you have found difficult
- take the waiting out of wanting – anticipate her feelings and volunteer the solace that you can see she needs
- take the opportunity, at a neutral, stress-free time, to tell her openly that your home will always be her refuge if she needs it

Teachers

- some girls can find it hard to admit errors and may get themselves into increasing trouble by offering multiple and increasingly thin excuses; try to intervene and forgive before a student digs herself in too deep
- if a pupil relies on school for her solace, make sure it's on offer somewhere
- help everyone, through class discussion, to be aware of this need, of who they may go to, and when and where, and why it can be important to seek refuge
- though girls are often open with friends, some may be friendless; peer counselling can make girls more willing to seek refuge within school

10 Make her feel she belongs

Human beings have a profound need to feel they belong somewhere and to someone. Your daughter's first need will be to feel loved by the two people who made her or who are responsible for her but, as she grows. the more friends, groups and institutions she feels a bond with and can identify with, the deeper will be her sense of self. When she fits in somewhere, it tells her something about who she is, that she's not alone and not a freak. Belonging to a family, a social or ethnic group, a club, school, or place of worship also signifies she is wanted, accepted and acceptable. It provides her with some guidelines for who she is and how she should behave.

If a girl grows up without any sense of belonging – to a family, school, etc. – if she feels rejected through, for example, heavy criticism, she is likely to seek acceptance and a sense of membership elsewhere. She may seek others who have opted out of trying to please, and instead gain pleasure and status in dangerous and illegitimate ways.

Parents

- tell your daughter family stories, so that she knows her own and your roots
- include her in as many family events as possible
- understand how fashion and uniforms can be symbols of belonging, especially for younger girls, and help her to 'fit in' – provided she does not become dependent, and that her needs can be accommodated by your family budget
- be on the look-out for signs of 'aloneness'; suggest that your daughter joins a sports or social club or youth group if she spends too much time on her own

Teachers

- circle time and similar arrangements can reinforce group identity and make each child feel an equal member of the class
- stable groups allow a clear identity to form; staff and student changes and re-groupings can be minimised for girls who may be especially vulnerable; at secondary level, pastoral continuity during the early years is crucial
- a class group with a high turnover of students throughout the year (known as 'turbulence') will need constant efforts to re-establish the group's coherence
- girls who have attended several schools will be especially needy

11 Allow her some privacy

My mother wanted to know everything about me, especially how I thought about things. It drove me mad and I felt sort of invaded. One day I screamed, 'Stop trying to get inside my head!'

As girls grow up, they like to mark their growing sense of independence and separateness by acquiring space and time to keep to and for themselves. Parents should not get unduly upset when their daughter tries to mark out territory that belongs exclusively to her, from which they seem excluded. This territory might be her bedroom, it might be a 'secret society' with friends, or it might be a diary. Later, it might be her social or her sexual life. The more appropriate personal and private territory she is gradually allowed as she grows up, the less likely she is to take over and lock you out of areas to which you need to maintain access.

If a girl's need for privacy and autonomy is not met, and it becomes a craving, she may be drawn into the manipulative, secret and life-threatening process of controlling what she eats.

Parents

- respect your daughter's need for some aspects of her life to belong to her alone; from quite a young age, you can give her a drawer, shelf or part of the garden for her sole use

- some children who feel 'invaded' or controlled by their parents will create a private world that involves a great deal of fantasy or secrecy, to the point of habitually stealing or lying

- never delve into her personal diary; if you suspect behaviour that may require your intervention, ask her straight out

- if food or school become her private space, try to give her more privacy elsewhere – in her head, in her room, or time after school

Teachers

- school is a very crowded and public place in which both action and participation are valued highly; girls whose home space is confined and similarly crowded may need to seek their privacy within school

- to respect and meet this need, offer quiet rooms or, for younger children, a quiet corner in the classroom

- girls who don't participate in certain lessons may be taking private time

- be watchful that girls who genuinely need quiet time do not become progressively withdrawn

12 Encourage a self-regarding femininity

Despite the fact that girls today have opportunities that were only dreamed of by previous generations, there remain a heritage and history of suppression, denigration and self-denigration that do not easily disappear. Men and women, and boys and girls, may indeed be different, physiologically, psychologically and hormonally, but that does not, and should not, translate into role rigidity or fixed assumptions about status and worth.

Girls are entitled to grow up feeling proud, not ashamed, of who they are. They should be able to look at themselves in the mirror and not only like who they see, because they know they count for something to somebody, but also feel excited at the prospect of exploring whatever talents they possess. Femininity is a given laid down by genes; it is not qualified by either looks or body shape.

Every girl must be encouraged to have a regard for herself that encompasses her gender but also respects her autonomy and the potential this gives her to influence her world.

Parents

- bolster your daughter's integrity; help her to be honest with herself and you, to trust her own judgement and to have ideals and live by them
- reinforce her individuality, not her inclination to follow the herd
- encourage her to care for others, but not to deny her own needs in the process or to define herself solely as a 'giver'
- try not to belittle 'womanhood', or any mother who chooses either to stay at home or to work full-time; allow your daughter to choose for herself
- don't tolerate 'bitchy', hurtful talk or cliquey behaviour simply because you think 'girls will be girls'

Teachers

- encourage a whole-school policy of zero tolerance of hostile or abusive talk, whether it be female-style bitching or macho-style, violent or aggressive language
- avoid time-tabling community service or environmental activities as an alternative to sport or anything military; no girl or boy should have to choose between being 'macho' or contributing to the community
- ensure plenty of class discussion to raise awareness of caring and gender issues

13 Support her when she's under stress

Contrary to common belief, stress is not something suffered only by adults. In fact, children get a double dose, from events in their own lives, such as bullying, academic pressure or friendship problems, and the knock-on effect of adult stress when parents become preoccupied and less tolerant. Given that children have less experience of life and of themselves to draw on to trust that 'normality' will resume, they are likely to be more confused and disorientated by stress than an adult, not less.

Events that may destabilise a child include separations from those close to her, including friends and pets; anything that will change the way she sees herself or how she imagines other people see her; and changes to either routines or relationships that upset familiar patterns.

If your daughter becomes sad or unhappy for longer than you would expect, becomes aggressive, withdrawn or socially isolated, sleeps badly, develops stomach and other pains or a strong thirst, loses weight or concentration, or becomes more dependent, she may be distressed and you should take action.

Parents

- if your daughter seems distressed, spend more time with her, and ensure that she gets plenty of sleep
- take her worries seriously, even if they seem insignificant to you
- keep her informed about changes and decisions, so that she feels less out of control, and keep regular routines going to shore up her security
- giving emotional support is tiring; get more rest and take some breaks so that you can continue to give what it takes
- remember, stress undermines self-esteem

Teachers

- look out for signs of stress, e.g.: deteriorating work; changes in behaviour; visits to the nurse; becoming more isolated, weepy and sensitive to criticism; leaving lessons to retrieve books or visit the toilet; someone destroying their own work
- find the time to talk to a student if you are concerned about her
- share your worries with relevant colleagues and contact the girl's home if your concerns continue
- be aware that your own stress may make you more vulnerable to and less tolerant of challenge from students or colleagues

CHAPTER 3

Deepening Her Self-Knowledge And Self-Awareness

If your daughter is to be able to stand up for herself, she has to have a clear concept of her 'self' – that is, who she is and how she feels. For too long, and in too many cultures, the ideal woman has been somebody who practises self-denial, in one or both senses: either she denies herself things occasionally, in terms of consumption or fulfilment, in order to meet the needs of other people; or, in a more extreme form, she is denied any self-expression or identity, in which case her notions of who she is are unclear. If she spends her life deferring first to her parents and brothers, and then to her husband, a girl will have merely a reflected sense of self that is as intangible as a mirage. To grow up with reliable self-esteem – with a self-generated, concrete, reliable and resilient sense of self-worth – your daughter will need to have not only plenty of opportunities to define herself and discover her potential, but also to be acknowledged and treated as an individual worthy of respect.

Before babies learn to talk and think, they know themselves only through their feelings and, throughout life, our perceptions and passions continue to be important to self-understanding. The fact that certain things frustrate us, make us happy, interest us, excite, upset or hurt us, defines our personality. Many parents find it easy to let their daughters continue to express their feelings as freely, but not all. Any child who is forced to cut herself off from her instincts and intuition must discard a part of her inner self, which can create a void. If her feelings are constantly denied and distorted, she may find it easier to ignore her own feelings and passions as a child, and harder to empathise with others as an adult.

Self-awareness and self-knowledge are also very practical attributes that contribute to motivation and learning. To progress, a girl has to know what she is capable of and what she still needs to know in order to master a skill or reach an objective. Self-knowledge and reflection are key prerequisites for self-direction. On an inspirational and intuitive level, it is the ability to look inside and know herself that will give your daughter a sense of awe and wonder, an appreciation of beauty and an understanding of the inner world of other people, on which lasting and intimate relationships depend.

The American play therapist, Virginia Axline, has written: 'The child must first learn self-respect and a sense of dignity that grows out of his increasing self-understanding before he can learn to respect the personalities and rights and differences of others.' In order to be sensitive to other people's sensitivities, we first must understand ourselves. Those who live and work with girls have a duty, then, to develop the girls' sense of self by heightening their self-awareness and self-knowledge.

14 Offer choices

As adults, we are fortunate to have a great deal of choice. It gives us more control over our lives. Common sense tells us that children should also have choices, but how much, when and, just as important, why?

In a world of multiple and unfamiliar choices, girls need to be able to make informed and responsible decisions. Your daughter will do this better if she understands her own preferences, is sure enough about them to resist outside pressure, and can think through the consequences of her choices on herself and others. Any adult who lives and works with girls should develop these thinking and reflective skills wherever possible.

Choice is important because it gives children some scope to influence what happens to them and stops them feeling powerless and put upon. By helping them to define what they like and want, choice also helps to sharpen their sense of self.

Parents

- younger children can be given appropriate choices about what they wear, what they play, who they play with and what story they have at bedtime
- 'how about playing with...?' – avoid answering the question before they do
- older girls can have some choice about when and where they do their homework, what (though not how much) TV they watch, what they spend their pocket money on, and so on

Teachers

- respect a student's decisions; don't ask her what she wants, then ignore her reply
- choices help to manage behaviour in the classroom; for instance, you can say: 'You can carry on talking and messing about, or you can have a detention; it's your choice'

15 Manage choices

In larger families, meeting everyone's whims and wishes can be impossible. It's not good for them, or you, in any case. Too much choice can, perhaps perversely, undermine a girl's sense of self: never having to make true choices, she won't discover what she really likes best. She can become confused and unhappy with too many choices about too many things and she won't feel her parent is taking overall responsibility.

Too much choice has other down sides: it may encourage your daughter to control and manipulate situations; it won't help her to live with the reality of disappointment; it may encourage her to become selfish and insensitive to other people's needs; and it can take responsibility from her because she can always say, 'Sorry, wrong choice – I'll have this one instead' if she doesn't like the consequences of her decision. Endless choice can also lead to rows: if all her wants are quickly satisfied, she may push and push to the point where you explode.

To help a girl strengthen and deepen her self-knowledge and self-esteem, the choices offered must be both limited and managed.

Parents

- managed choice means 'either/or' decisions; you put limits on the choices, having already decided what you are happy to agree to
- limited choice means keeping these choices to a few times a day
- avoid open-ended choices; on a cold day, it's better to say to a young child: 'Would you like to wear your jeans or your track suit?' in case she chooses something unsuitable; and only two or three items should be on the breakfast menu
- make sure the choices you offer are ones you can make happen
- girls should not normally be in charge of how the whole family spends its time

Teachers

- choice is motivating; students who are given some choice about what they do and how they do it are usually more committed to their work
- include aspects of choice within project work – not so much that it becomes hard to start, but enough to let a student make it her own
- where there is little scope for choice in lesson work, activities that focus on choice can be tried, such as: if 'I could be a food/tree/colour/musical instrument/car/country/piece of furniture/an animal, I'd be a ..., because ...'

16 Don't impose your views on her

My father was very autocratic. He couldn't discuss anything, he just stated his view and declared all others to be uninformed and stupid. He tried to tell me what to think and how to do everything. I ended up taking no notice of anything he said.

It is very easy to become so convinced by your own status and wisdom as a parent that you dish out declarations, decisions and assertions without realising it, squashing your daughter's growing need to explore her own views on any matter. Too much criticism or praise, or excessive use of rewards, can do the same; you are asking her to live by your views and values.

Pre-teenage girls are considered to be 'biddable'; they like to please and conform. You should not take advantage of this, and should make a special effort to foster your daughter's individuality and self-awareness. Later, when girls define what they're not before they explore who they are, it is very common for them to reject their parents' values. They should be free to cross that road without being run over.

Parents

- seek your daughter's views; say, for example: 'I like it, but it's what you think that's important', not 'That's great, don't you think?'; and 'What did you make of that TV programme?', not 'That programme was a load of rubbish'

- think carefully about the things you care about, and be aware that your daughter is likely to target these as she asserts her independence; if she does, don't take it personally

- even with a younger child, the more you push your views and values and assume she should share them, the more likely she is to reject them

Teachers

- encourage students to think ahead about what might happen next – for example, in a science experiment – rather than telling them what they are about to see

- resist the temptation to save time by delivering the standard arguments for and against something in class work; always seek students' views, to develop their confidence and thinking skills

- if you get involved in sorting out a conflict between two girls, don't impose solutions on them but encourage them to devise their own

- in general debates, keep your views to yourself to allow students to explore theirs; this does not mean you cannot question and challenge gently

17 Give her feelings space in your world

As a girl, I was forced to live on an emotional plateau, never allowed to express boundless joy or the depths of sadness. Being constrained by moderation in all things suffocated me and I almost lost myself. My own daughter skips about when she's happy, and I love to see it.

If someone said to you, 'You've no right to feel that!', you would probably explode. Children also get upset when their feelings are denied. We now understand that feelings are as important as thoughts in the development of our children as unique and giving human beings. If we reject our daughter's feelings, we reject her as she experiences herself.

Feelings used to be viewed as somehow inferior to thoughts, more closely associated with instinct and animals, something that governed us when we were primitive, not civilised and sophisticated. We now know that they have an important role to play in tandem with conscious thought, partly to assist survival when life is threatened, and partly as a way to understand the situations in which we find ourselves.

Parents

- if you can accept your daughter's feelings, she can learn to live with, manage, enjoy or work through them herself

- help her to ask for what she needs; 'I think you're feeling upset. Would a hug help?' can free her to say, 'I'm feeling down, so I've come for some comfort'

- you accept her feelings when you accept her apologies; you could say something according to this model: 'I was sharp with you because I had an awful day today. I'm sorry'

- feelings of jealousy, anger, frustration and resentment should be accepted, not punished or denied; however, while it's fine to feel, it's not fine to hurt someone because of the way you feel

Teachers

- give girls sentences to complete, such as: 'I'm happiest when...'; 'When I get angry, I...'; 'I feel most important when...'; 'I feel frustrated when...'; 'I tend to give up when...'; 'When I'm told off, I want to...'

- brainstorm emotions: invite the class to discuss how they feel about or see something; explore words which express emotions, e.g., frustration/ powerlessness/ anger /rage; students can discuss trigger events, and decide whether the feeling anyone describes fits the one under discussion

- give younger children a 'feelings' log book, in which they write, at set (and free) times, their reactions to pieces of work, school, events or people

18 Tell her her story

Once when you were little, you got so angry that you hid behind the sofa with some scissors and cut a hole in it! Then it was my turn to be angry!

Young children love to be told stories about themselves – when they were babies, how any older brothers and sisters reacted when they were born, and so on. These anecdotes give your daughter a history. They are the jigsaw pieces of her life that she can't reach and which she needs to complete the picture of herself.

Older children like to hear different stories – about their parents' childhood and school days, or the antics of any aunts and uncles. Such tales will deepen a girl's sense of belonging because she will understand better what it is she belongs to. Each story will act as a connecting thread that creates continuity. Like a spider's web, the more threads she has, the stronger she will feel.

Difficult times need to be talked about too. There will be gaps in her history if particular incidents are blanked out.

Parents

- get out the family photos occasionally; talk about the people and events shown in them; this can fill in gaps in his understanding of family history, generate laughter, lead to other linked issues, reinforce your daughter's identity and increase her confidence in the future
- regularly recall past holidays, birthday parties, special treats or school outings that were fun and brought the family together
- if you can, keep mementoes such as your daughter's first shoes, favourite toys, books and clothes, so that you may revisit the past and show it's dear to you
- discuss difficulties, don't bury them; she has a right to information about herself so that she can make sense of her world

Teachers

- personal life-lines: in a group, discuss the many different experiences that have featured in pupils' lives, what made it significant and how they felt at the time; ask each child to draw a vertical line on a large sheet of paper; the line represents their life from birth to date; they then write in their own personal events, positive and negative, on either side of the line
- not everyone has a happy family story to tell; focus on both good and bad, happy and sad experiences, to ensure a full and realistic picture that leaves no one out
- make sure you include all types of family and caring arrangements when you discuss family matters with your class

19 Encourage the practice of reflection

Reflection means standing outside oneself, looking in at what has happened and asking questions like 'Why?', 'How?' and 'What if?' To reflect means 'consult with oneself, go back in thought'. Reflection entails reliving your experiences and having a silent conversation with yourself to identify connections and understand them. It increases self-awareness, and helps you to make sense of your life, take control and introduce changes.

Reflection is beneficial because thinking backwards is the first step to thinking forwards, which is important for learning, sustaining relationships, managing conflict without violence and for planning one's life. If a girl cannot reflect on what she has said and done, assess the good and the bad and realise what she might need to change, she cannot develop and make progress in anything. Becoming aware of how she feels and behaves helps her to understand other people's feelings and actions, and therefore to anticipate possible problems in relationships. If she can understand her past, it will help her to face the future.

Parents

- recap and reflect on the day's events with your daughter each evening as part of the nightly routine, and invite her to 'think aloud' about both good and bad aspects of it
- do this yourself; say to her: 'I wonder if I could have done that another way?' or 'I really felt excited/angry when it happened', which demonstrate reflection in action
- ensure that your daughter reads fiction; where possible, discuss the events and characters the stories contain
- encourage drama, imaginative play and dressing up, depending on her age

Teachers

- ensure that girls read fiction regularly as well as non-fiction, and discuss the story line and characters
- introduce role reversal in drama to explore alternative experiences
- ask each child to write down two negative and two positive descriptions of herself on two pieces of paper; transfer them to a large board; ask, for example, 'are you always irritating? If not, when are you, and why?' to help girls reflect upon and reject negative labels which they permanently assume

20 Arrange relaxation and quiet time

Most girls have times when they like to be on the go and let off steam, but every girl also needs to be able to slow down when the situation or a person requires it and to feel comfortable within herself when she does stop. It's good to be active, but being bored isn't a crime. Children are more likely to explore and develop their inner selves when they are quiet, and should not panic at the prospect.

Quiet time means the chance to be calm, to wind down, relax or even escape. Quiet time also means being peaceful, at rest and happy to be alone. Each girl has her own way to unwind. One will prefer to flop in front of the television. Another will choose to play with a pet. Some like to relax while listening to music; others will be happy to reflect as they draw or doodle.

Being quiet at these times gives a girl the chance to let her thoughts roam, to find out what is inside her and learn to be content with and within herself.

Parents

- discover your daughter's mechanisms for winding down and encourage her to use them, especially after school or when she's been very active – times when she may find it hard to 'come down'

- respect the older girls' need for their own quiet space, which may not be within your home if it is a very busy place

- encourage quiet togetherness – watch TV or a video with your daughter, make time for morning cuddles in bed if she is still young enough to enjoy them, or sometimes drive her to her friends' homes

Teachers

- simple breathing and relaxation exercises can work very well with primary school children; older girls may need more encouragement, and may find yoga or the Alexander technique more acceptable

- have a clear ending planned for lessons, especially those in which there has been a lot of activity; use this time to recap, and encourage pupils to reflect on what they have learnt

21 Explain your thoughts and feelings

'Children need models more than they need critics' - so said the French philosopher, Joubert. Your daughter will learn to identify and express her thoughts and feelings, safely and comfortably, if she sees you doing so.

One mistaken but common view of assertiveness is that thoughts and views should be presented in a commanding and dictatorial fashion as 'the truth'. This denies the possibility that others, including children, may see things differently, and it may lead to conflict.

Thoughts and arguments should be proffered and explained, not asserted, with the possible exception of key matters of discipline. A useful principle is: assert your right to be heard, not your view of the world. Inner strength is based on tolerance and respect, not domination, and taking this stance is also hugely liberating: if you don't need to dominate, you don't need to be right each and every time.

Parents

- if you are angry, upset or frustrated, say so and explain why; don't just shout
- when you explain your reactions to your daughter's actions, it helps her learn to anticipate, predict and become more thoughtful, sensitive and responsible for her behaviour
- tell her how you arrived at a particular decision, by saying, e.g.: 'I first thought this, then I realised that, so I decided...'
- you alone are responsible for your feelings; say: 'I felt angry when...', not 'You made me angry'
- if your daughter swears, ask her to find an alternative word to express what she is trying to say, and do the same yourself

Teachers

- help your students find the best language with which to express their thoughts and feelings
- be a model for this yourself, and endeavour to express your own thoughts appropriately
- use phrases that start with 'I' to avoid appearing to blame anyone
- emotional literacy requires a particular vocabulary; girls need to discover which words they can use to express their feelings – and, if information is presented properly to them, will enjoy widening their vocabularies to replace expletives

22 Maintain communication

I thought my daughter was fine. She seemed to get on with her life, though she was alone a lot, and I got on with mine. We didn't talk much. Then, in the middle of her GCSE exams, she withdrew from everything. Now she won't work, go out or talk. We're in a real mess.

If children and adults don't communicate, the price to pay can be very high indeed. Communication skills lie at the heart of social and emotional health and success, and a girl will not be as comfortable about talking if adults, especially parents, don't talk to her. No conversation implies no interest, which she may interpret as neglect, so family silence can have a devastating impact on her self-esteem, her sense of self-worth and her trust in future relationships.

But family conversations will also help your daughter to be comfortable about expressing her views in front of other adults in positions of authority as well as her peers. It will increase her confidence and help her to stand her ground with professionals and officialdom in the future.

Parents

- keep talking, even if it feels uncomfortable; the more you do it, the easier it will become
- try to eat with your daughter as often as possible; if this happens only occasionally, forget table manners and other controversial subjects; swallow hard if she challenges you, and don't rise to the bait
- if you want to take issue with something, start with the word 'I'; by saying, for example: 'I'm not happy about doing all the housework and I'd like some help'
- always stop and listen
- if you have to be away, for work or for pleasure, give her your undivided time on your return to touch base with her again

Teachers

- withdrawn or silent students reduce demands on teachers, but it must never be assumed that they are coping perfectly well
- although quiet students may simply be taking time out, be vigilant; identify any longer-term patterns; consult your colleagues, talk informally and socially to such students, and step in sooner rather than later if you are still concerned
- ensure that lesson plans entail a variety of exercises that encourage involvement, and include work in small groups
- in all lessons, nurture the essential skills of communication: reflection, listening and tolerance for others

23 Encourage self-assessment

Although our children like to know they have pleased us, what we should try to encourage in them from the start is the confidence to evaluate and praise themselves.

Girls are inclined to undervalue their work and worth and are typically self-effacing. For some, this may help them to put in the extra effort they believe is required to improve their results, but this can reinforce the view that 'I only did well because I worked – I'm not really that good'. However, with support and guidance, they can learn to assess themselves and their skills realistically and to do themselves justice. The particular danger time is the early teens. Girls' self-esteem dips noticeably around the age of 14.

Scepticism about praise and the sting of criticism are removed when the evaluation comes from themselves. Self-assessment is important because it is central to independent learning, which is the way of the future.

Parents

- avoid being generally critical and judgemental so that your daughter does not come to depend on your opinion and lose faith in her own
- when your daughter asks what you think of something she has done – a painting, an essay, an achievement, a music practice – turn the question back to her; what matters, ultimately, is what she thinks of her effort, and encouraging her to trust herself is crucial
- minimise criticism, especially in relation to a girl's looks and abilities, for it will force her increasingly to seek approval from you

Teachers

- girls can be encouraged to assess each other's work, in pairs, as a first step to learning how to be honest about their own work
- if pre-school pupils can work well with a 'plan, do, then review' approach, older girls can certainly manage this too
- full and accurate feedback is vital; at the end of any work, a student can be required to assess its merit, then you can detail why and how it did, or did not, meet the standards for that assessment
- once self-assessment becomes frequent and normal, there will be no excuse for a girl to underestimate her abilities

24 Hear the sound of silence

Between fourteen and sixteen, our daughter hardly spoke to us, except for yelling! She spent hours in her room and became almost a recluse. I was really worried. Then her skin started to clear and, a bit like a butterfly, she emerged more confident and began to socialise again.

In the end, this parent trusted her instinct and everything turned out well; but take care. If your daughter withdraws and stops talking, it doesn't necessarily mean you can assume she's all right, because she might not be. It could be a sign of unhappiness and depression.

However, your daughter does not have to talk all the time. Silence can mean she feels comfortable and does not need to fill every moment with words; it can be understood as quiet togetherness. The important thing is not to ignore it if a girl goes quiet. Hear the silence, reflect on it, accept it for a time, see if you can find out what she is doing when she is on her own. And step in if you believe she is withdrawn from others as well as you, and if there are additional signs of problems.

Parents

- keep talking – about things that won't start an argument – but don't force your daughter to respond
- suggest you do something together that doesn't rely on talking but that you can share, like swimming, bowling, or going to the cinema
- be aware of signs of difficulty and danger, such as leaving late for or missing school frequently, changes in her eating patterns, unusual smells that could indicate lone drug or alcohol use, or any drop in the standard of her personal care

Teachers

- small-group activities encourage participation and offer less chance to hide behind silence
- a girl who clams up in the classroom may be afraid of making mistakes or be distracted by problems; it's important to find out why she isn't talking
- try to include activities that require everyone to participate

CHAPTER 4

Giving Her a Positive View of Herself

The twenty-first century is likely to be characterised by considerable change and uncertainty. To manage this successfully, a girl must see herself and her capabilities in a positive light. This does not mean she must see herself as perfect – indeed, it is better if she does not.

If your daughter feels sure of herself, she will find it much easier to resist, or at least to manage sensibly, undue academic pressure, peer coercion and the temptation to use alcohol, drugs and sex inappropriately. As an adult, her 'can do' attitude will help her to stand up for herself, and she will be better able to manage her finances, relationships, children and career. However, if she sees herself as a source of problems, mistakes, disappointment, pain and distress, she will lack the confidence to take on challenges or commitments.

So how can adults help to create 'can do' girls who are motivated, enthusiastic and full of optimism? It is vital to demonstrate that your daughter is important to you and to remain supportive, give plenty of positive feedback, extend her horizons and do all the things that will make her feel loved and wanted, but it is also crucial that you minimise the blame, nagging and fault-finding that convey disappointment. Without intending to, you can send harmful messages about how likeable and competent she is, with damaging consequences for her mental health, stability, self-esteem and motivation.

Constant criticism will make her feel she can never please, that there's something fundamentally wrong with her. She will always be looking over her shoulder, wondering which of her actions will be next for disapproval. This undermines confidence, initiative and morale, as do shouting, unwarranted blame and routine harsh, erratic punishment.

We find many excuses for our damaging words, and may say that she deserved it. We may regard any challenging reactions from her as a sign that the words did not hurt – not realising that the shield she raises is a protective fiction. If she reacts with hostility, we often see it as rejection, and think why should we bother to be pleasant to her anyway? If we sense any hurt, we may tell ourselves it's time she grew up. But a girl who 'cannot take' constructive criticism has often taken a bucketful of it and should be given no more if she is to have any energy left to protect her self-respect.

Neglect hurts, too. When parents lead lives in which their daughter hardly features, or give her more freedom than is right for her age, she may interpret this as indifference. What a girl needs is not her parent's declaration that she is loved, but the demonstration of this love, through action and words, in a manner that she can appreciate.

25 Understand her particularities

Every girl is different. She will feel things differently, play differently, think differently, learn differently, and enjoy different things. It is these 'particularities' which define who she is as a human being, regardless of her gender. A girl will have a clearer appreciation of the different elements within herself if the adults who know her well put what they see into words and encourage her to do the same.

To take the analogy of an artist's palette, the more 'colours' or traits and talents that are identified, the more interesting and colourful the picture that can be painted. Too often, parents see their daughters in black and white, not in colour. They are either 'good' and 'successful', or 'difficult' and 'hopeless'. To remain proud, happy and confident, girls need to view themselves as multi-dimensional and multi-coloured, having a variety of positive personality traits and abilities.

Parents

- fill in your daughter's 'personality palette'; write down her likes and dislikes – what she likes to eat or won't eat, her favourite games, pastimes and activities, the clothes she likes, what she's good at, the places she likes to go, how she works best
- be positive: traits you view as negative are probably the reverse side of positive ones; for example, she may stand up for herself with friends, but be 'too assertive' with you
- tell her what you see, for instance: 'I really like the way you...', or 'You're very caring/sensitive/good with your hands, aren't you?'

Teachers

- consult with other staff members to determine specific strengths and weaknesses of any student you find 'difficult'
- be aware of different learning styles and vary lessons appropriately
- ask your students what learning styles they prefer
- ask students to get into small groups; select one student in turn, then have the rest of the group tell her all the strengths and unique features they see in her (no put-downs are allowed); one person should record the contributions, listing ten to fifteen strengths for each person

26 Don't compare her to others

For some girls, living in the shadow of a brother or sister is a nightmare that persists for the rest of their lives. A girl may never quite shake off the humiliation and feeling of inferiority. Comments such as 'Your brother would not have produced work like that' or 'You're not as talented as your sister' can ruin pride and kill ambition. Friends, too, can be used (or rather abused) as a model for her to measure up to, in questions like: 'Why can't you dress smartly, like Riya?' These taunts, far from acting as spurs, tend to taint the childhood of those who suffer them. Comparisons undermine confidence and fuel conflict.

Even handing out equal praise can be limiting. Saying: 'She's the brainy one in the family and he's the athlete' may give each child something to be proud of, but will also make it less likely that either will explore their potential in the other's field of interest. Although brothers and sisters may genuinely be different, they make themselves different to create their own territory. When skills become territories, children can become tribal. And don't forget that comparisons with any other girl's body shape or weight should be totally banned.

Parents

- celebrate difference; every child is entitled to be different, look different and respond differently, because each one is unique
- make it clear that there is room for more than one artist, pianist or athlete in the family
- discourage any family focus on body image or shape, and don't compare your daughter's body with anyone else's, favourably or unfavourably
- don't compare your daughter with how you were or what you did at her age; she is herself, not you

Teachers

- positively value each child as an individual; references to siblings should never be used to disparage or coerce work
- be especially supportive of originality and creativity, because these are the manifestations of a girl's unique 'self'
- competition promotes anxiety; if comparisons are to be made, the most constructive are with any individual's last performance or piece of work

27 Respect her feelings

Feelings are fundamental; they make us who we are. Although in general parents will find feelings such as fear and anxiety easier to accept in girls than in boys, there is a danger that the pursuit of equality and success will tempt parents to restrain such feelings in their daughters as well as their sons. There is another reason many adults would rather even girls stopped being 'emotional': the sooner children can control their fears and feelings, the sooner parents can stop having to 'waste' their time responding to them. Fear of the dark, of water, spiders, losing friendships, fear of failure, of nightmares and bogeymen, test parents' patience; they try to respond with rational arguments to a fear that is actually irrational. For your daughter, the fear may not be on the rational spectrum – it may instead be purely emotional, and the two don't mix.

Whether they feel delight or disappointment, fear or fury, joy or jealousy, girls are entitled to have their emotions acknowledged and respected by their parents and carers, because this is the route through which they experience the essence of what is themselves.

Parents

- respect your daughter's fears and anxieties, hopes and dreams, even (and especially) when they seem silly or irrational to you
- share her delights and disappointments
- acknowledge and describe how she might be feeling, so that she develops a language that will help her to understand her emotions

Teachers

- fear of failure explains a wide spectrum of behaviour that obstructs learning; encourage students to be open about their fears
- at all ages, drama and role-play can allow both girls and boys to explore emotions 'safely'
- some girls are naturally intuitive and aware; group debate and discussion will help those with less insight to learn from those with more
- reading fiction is an effective way to explore experiences and consider their impact

28 Listen with both eyes

My dad doesn't really listen to me. I'd love him, one day, to put his newspaper down or turn off the TV when I'm talking to him. I feel like a nobody.

Listening involves looking just as much as hearing. When you listen with half an ear, it usually means you are concentrating on something else you are doing, not looking at your daughter. When you look at her, three things happen: first, you have to stop what it is you are doing, so that your full attention is available to be directed at her; second, eye contact with her helps you to fix your thoughts on her and what she has to tell you; third, you are able to read her 'body language' and facial expressions, which will help you to understand anything that lies behind the words.

When parents and teachers don't listen properly, or at all, they are saying, in effect, that they and their business are more important. If a girl is ignored, she will feel insignificant and undermined. She will not feel comfortable with herself and her self-esteem cannot possibly flourish.

Parents

- when your daughter wants to tell you something, use your eyes first, not your ears; stop what you are doing and focus on her
- look at her facial expression and watch for any hidden meaning; notice how she is standing or sitting and her tone of voice
- let her know you are taking her seriously, and say: 'This sounds important; I think I'd better sit down and listen to you properly'

Teachers

- show awareness, and say: 'My antennae tell me there is more to this story than you are telling me'
- teach students about body language and be vigilant about its application in all oral activities in the classroom
- provide plenty of opportunity for girls in discussions, debates and presentations to listen to and respect each other's contributions

29 See with both ears

Like listening with both eyes, seeing with both ears is about helping adults to be more sensitive to a girl's inner world, which is as important to her as the outer, more visible, one. Both influence the quality of her self-esteem, the former being more important.

You get a glimpse of what might be happening inside your daughter by listening to her 'self-talk' – what she expresses about herself. Adolescent girls are particularly prone to run themselves down. For example, a girl may seem to be doing well at school, and to have distinguished herself in a particular test. However, she may respond to the result by saying, 'That was a fluke. I didn't deserve it' or 'I'll probably fail next time.' Her words show that, inside, she doubts herself. Similarly, she may have plenty of friends, but if one cries off a visit and she retorts, 'She's probably had a better invitation,' it indicates a tendency to belittle herself with negative self-talk.

Parents

- listen for times when she talks ill of herself, and rephrase her comments to make them accurate and/or positive

- try to keep a record of what your daughter says and how often she makes self-deprecating comments, even if in apparent jest, to help you understand any pattern or the scale of the problem

- simply denying a child's self-criticism won't have much impact; turn a generalisation into a specific, so that 'I'm never any good!' becomes 'You did not do as well as you wanted this time'; or repeat several times that what you see is different: 'I find you quick to see the point/amusing and fun to be with' or 'I see you as someone who...'

Teachers

- be positive; discourage negativity and challenge 'I can't do this' assertions

- encourage self-evaluation tasks in which students write about their performance – and highlight areas where they believe they performed best

- if a girl says she's no good and knows nothing, draw a horizontal line with 'knowing nothing' at one end and 'knowing everything' at the other; invite her to mark the appropriate place that represents how much she really knows; she'll realise that she certainly knows something

30 Respect her play

As a child, I had a brilliant time, making up stories, getting up to minor mischief, spending hours outside with my friends.

Fortunately, girls love to play: play is essential to the development of self-esteem and confidence. They are particularly good at developing story lines as they 'make believe' and 'play act' with their friends. Through play, girls find out who they are, because through the choices they make about what to play or do, who to play with, what to draw, and so on, they gradually flesh out their idea of who they are and gain an identity – two essential steps in gaining self-esteem. Through play, girls also discover what they can do, because play develops verbal, social, manual, planning, problem-solving, negotiation and physical skills, which enhance their self-confidence and their ability to socialise and make friends.

Finally, through safely managed, independent play, girls gradually realise that they can manage on their own.

Parents

- encourage your daughter to play, with you, with her friends, on her own, indoors and outdoors
- let her choose what to play most of the time
- respect her play by giving her notice of when she must stop, and don't spoil her fantasies by teasing or ridiculing her
- pretending, dressing up, drawing and creative games are all important because they help girls to become spontaneous, imaginative and creative, which helps them to do well at school
- help your daughter to become familiar with computers, for they are the way of the future

Teachers

- show respect for students' hobbies and interests, and utilise these for individual presentations and projects
- role-play and improvisation can be great fun and can provide a valuable release for girls
- though play is important, there are places and times when it is appropriate and places and times when the fun must stop; building in 'winding-down time' after intense activity can help to mark the boundary between the two
- girls should be encouraged to extend their play repertoire to include action games as well as quiet activities

31 Let her impress you

All children get a real boost when someone they admire a great deal is clearly impressed by something they have done. They puff up with extra confidence and pleasure. It does their self-esteem no end of good. Showing that you are impressed is, of course, a form of praise, and a very effective one, which avoids some of the pitfalls associated with praise.

Being impressed is straightforward, open, and takes away some of the measured judgement implicit in other forms of praise. Most important, it is unconditional. 'I am really impressed!' or 'That was impressive!' says it all – no ifs or buts to qualify it or detract from the core message. There is nothing grudging about being impressed.

To say you are impressed also clears the air of competition. Some parents feel they ought to be stronger, better and more successful than their offspring and need to prove it. But being impressed by your daughter doesn't mean that she has got the better of you, and it won't stop her trying; it simply shows admiration, which is what girls thirst to receive.

Parents

- involve your daughter in real tasks, working alongside you; ask her to help you mend things, sort things, decide things, clean things, then say, 'Wow, you were good. I'm impressed!'
- show that you respect her skills and her views: 'You're good at mending things. Can you help me with this?'
- give her appropriate responsibilities, so that she can test herself, develop her skills and feel trustworthy – then let her know she has achieved this
- when she's little, let her win in little ways, and show how impressed you are
- you can genuinely admire her computer skills, which probably will be far superior to yours

Teachers

- allocate responsibilities, such as reporting back from group discussions, looking up something that will aid everyone's classwork, and make it obvious that you are impressed by the outcome
- begin questions in the classroom with, for example: 'Gemma, you know quite a lot about this...'
- for an impressive piece of work, gauge the appropriate reward, public or private

32 Fashion her individuality

The media and images affect my life to the point where it gets silly... There are always people slimmer and nicer-looking than me and it knocks my self-esteem. – GIRLPOWER (See page 224)

Very few girls will end up having a model-shaped body or the money to clothe themselves in high fashion every day of the week. The current western cultural focus on body image means that just about every girl will possess some feature that fails to match her expectation or dream, and about which she may become depressed – the shape of her lips, the spread of her buttocks or the size of her eyes. Even primary school-aged girls can become obsessed with their appearance.

But looks are, literally, superficial – and not even skin-deep when they are cosmetically created. Today, the asset of attractiveness is valued disproportionately highly. If everyone in your family respects social more than physical qualities, accepts difference and shows respect regardless of physique or good looks, your daughter's self-esteem will have a chance to strengthen before it is undermined by her peers.

Parents

- teach your daughter to honour and care for her body by supporting her when she's ill and ensuring that she's clean and healthy
- don't reinforce fashion-induced stereotypes with sexist remarks
- beware of dressing her in fancy clothes, regardless of practicality
- if it is practicable, a regular, limited clothes allowance may help her to avoid being a slave to fashion, because she'll be forced to make practical choices
- help her to walk tall, literally and figuratively, because she's proud of who she is inside, regardless of her body shape and wardrobe

Teachers

- avoid making comments such as: 'You're nice and slim; you can be in the play'
- avoid team selection systems that leave the larger girls waiting until last
- ensure that pupils are aware of the power of advertising and the fashion and health industries to focus on body image
- talk about what is implied by different images of womanhood and manhood, and the role of individual choice and variety
- ensure that anyone teaching PSHE or nutrition is fully informed of the health needs of growing bodies

33 Lead her when she's ready

When I got to the sixth form, I had a sort of crisis. I'd been very active out of school, but I realised it had all been organised for me. I had to make choices then, to become an individual, and, because of my parents' involvement, I didn't know who I was or what I really wanted.

Parents can't wait for their children to walk, get out of nappies, read, swim, ride a bike and so on, for many different reasons, some more honourable than others, but this eagerness can undermine a girl's self-esteem.

Girls, like boys, learn best when they are ready. 'Ready' means not only being willing – hungry for the knowledge or skill – but also being comfortable and confident about moving forward. They should instinctively know that the necessary prior knowledge is in place to allow them to make sense of what they find. When a girl can influence what she does and when, she begins to know herself intimately and to trust her own judgement, which reinforces her self-esteem.

Parents

- 'If it's Wednesday, it must be ballet' – some girls are driven, literally and metaphorically, by parents to attend a stream of after-school activities, to keep them busy and give them opportunities to shine, but children get tired; beware forcing your daughter into extra-curricular pursuits which she doesn't enjoy and at which she's unlikely to succeed
- avoid pushing her to progress too soon; if you push too hard, she may actually lose ground

Teachers

- encourage 'mastery learning', in which students are given targets, guidance on how to proceed, shown assessment criteria and encouraged to manage themselves within these guidelines; mastery learning programmes are particularly effective with weaker students
- when a child is motivated, her work improves; work to enhance each individual's motivation, to encourage self-development
- girls are often happy to work on their own, but if a student seems to be 'stuck', a collaborative, group-work situation may help her to move on

34 Accept her friends

Friends boost my confidence, because they know me best.

Friendships are extremely important to girls. Friends help her feel she belongs and validate her, because they tell her she is liked and is likeable. They frequently share her interests, help her to fill her time, give her fun and an identity. She thinks: 'I am friends with this sort of person, so I am also like this.' Real friends offer loyalty and support when things go wrong. In other words, they can play a crucial part in building her identity, her confidence and social skills as well as providing safety in numbers, provided it is the 'right crowd'.

A girl's friends become so close that they are almost an extension of her; if you reject them, you reject her. This makes it extra hard if you want to suggest she's made the wrong choice. If her friends seem to subvert your plans, challenge your values and cause you concern, think very carefully before you try to exclude them. They may simply be normal adolescents needing to break away. Talk to your daughter before you forbid anything, for this may be counterproductive.

Parents

- invite your daughter's friends to your home or on outings with you so that you can get to know them better
- find something pleasant to say about them
- compliment your daughter on her ability to be a good friend to others
- if you're worried that her friends are a bad influence, try listing what she's getting from this group, and consider whether these needs can be met in another way; try also to talk about what she expects from a real friend – respect for her point of view and limits, wanting the best for her, and reliability, for example – then let her decide if her current 'friends' have these qualities

Teachers

- though natural friendships should be recognised and respected, you should arrange classroom seating and the composition of groups for project work in ways that encourage mixing and minimise peer pressure, bullying and isolation
- explore the issue of friendships and peer pressure through school assemblies and Drama; pose the question 'What makes a good friend?' and invite individuals to write down what qualities they believe they have as a friend to others
- work with a group to draw up a 'friendship contract': a list of behaviours one would expect from friends

35 Enter her world, carefully

My mum asked me what my favourite band was at the moment, so I told her and played her a few tracks from a CD. She listened, then screwed up her nose. She said sorry for not liking them but I didn't really want her to. It's my music.

It can be hard to judge the amount of involvement any girl wants a parent to have in her life, especially when she is at an age when she needs to become more separate and independent. There can be no clear answers – parents must simply remain sensitive to the issue and judge each situation as it arises. But there are two useful principles to bear in mind: take an interest (without being intrusive), and remember that your prime role is to be your daughter's parent rather than her friend. You can be effective and loving without being her best friend, who should be someone from within her peer group.

Parents

- take an interest in your daughter's hobbies and activities, but don't take them over; they don't have to become your passion too
- give her the space and territory to be separate and different, without cutting yourself off from her
- music and dance are frequently used by girls to explore and establish an individual identity; ask which band or style is her favourite, but don't make them your favourites too
- sporting events and matches can be shared safely and can bring different generations together

Teachers

- in any class discussions about personal and family matters, acknowledge your students' range of family types and personal experiences, but tread and talk very carefully in these areas
- think ahead about how you might respond if a student ever becomes distressed during discussions of personal issues

36 Keep criticism to a minimum

When I was a teenager, my mum said that, above the knee, my legs were better than my sister's, but below the knee, my sister's were better than mine. I was incensed that she could be that distant, disloyal and judgemental, even though her comments were supposedly even-handed. I never forgot this.

Adults are usually totally unaware of the destructive impact of their careless words, which can do untold damage. Even an occasional clumsy statement made in jest can allow self-doubt to take root, and both girls and boys are more sensitive to criticism than most adults realise. Constant criticism implants not only self-doubt but also guilt and shame about letting parents down. If a girl fails to please, she'll assume she disappoints; and eventually she'll feel useless and rejected, though she will probably hide it well.

In girls, this sense of shame can take dangerous routes, leading them to prove their worth through academic perfectionism, seeking the perfect body or even to self-harm if they feel utterly worthless.

Parents

- select one behaviour at a time and ignore the rest; piling on the criticism will make your daughter resentful and uncooperative
- accentuate the positive – focus on what you want done, and select one day when you comment only on the good things
- try to stop watching and judging, because this implies that you are controlling and mistrustful
- banish humiliating phrases like: 'I can't take you anywhere', 'I wish you'd never been born', and 'You make me sick'

Teachers

- teachers' words can hurt as much as anyone else's
- like criticism, teasing, sarcasm, ridicule, shouting and blame are put-downs which hurt, shame, degrade, damage and humiliate; they sap independence, initiative and morale and are never justified
- it takes four 'praises' to undo the harm of one destructive criticism
- turn all your 'don'ts' into 'do's'
- doubts are more cruel than the worst of truths: keep students' self-doubt at bay, and listen out for any tendency to self-criticise

CHAPTER 5

Demonstrating Care Through Love And Rules

We show that we love and care for our girls in many different ways. We make sure that our daughter has the right kind of food, so that she grows up strong and healthy. We ensure that she's clean and appropriately clothed, so that she is warm and protected from disease and ill health. We spend time with her, have fun and share our lives with her, which help her to feel loved and wanted. We try to understand and tolerate the mistakes she inevitably makes as she tests herself in the process of gaining skills, confidence and maturity.

However, giving in to your daughter because you can't be bothered to argue won't persuade or convince her that you really care; neither will showering her with presents. Having guidelines for behaviour in and outside the home that will protect her, you, others and even her physical environment demonstrates your concern for her future.

The right kind of discipline will nurture your daughter's self-esteem, because it will make her feel looked after. It will give her the freedom to explore and take risks within safe boundaries that you carefully define and manage. When she has clear guidelines for behaviour and a daily routine, she can relax. She won't have to decide everything for herself, or worry whether she may get into trouble. And when she does behave broadly as expected, life is not only calmer but filled with the warmth of other people's pleasure and approval. Most important, when the adults who look after her establish appropriate limits for a girl's behaviour and are sufficiently involved to notice and watch what she does, she realises that they care about her.

Of course, any rules must be fair and reasonable. The popular phrase 'tough love' does not give any adult the right to be brutal. The aim should always be discipline without dictatorship, and punishment without humiliation. When parents misuse the greater power they inevitably have through the imposition of harsh and humiliating rules and punishments, any child will feel affronted and try to get her own back. If this develops into a negative, tit-for-tat pattern, the constant retaliatory put-downs will progressively erode her belief in herself.

All children thrive on the approval of both a mother and father figure. If either of these arouses, instead, feelings of hatred and resentment through harshly administered discipline or indifference, a girl will react defensively to the perceived insult and create distance between herself and the very person she needs to feel close to.

By showing you care about your daughter, by providing love and (not too many but well-chosen and consistently applied) rules, you help her to care about herself and others, including you.

37 Love her for who she is, not who you want her to be

One of the hardest things to do is to love and accept a daughter for who she is. Instead, we dwell on what we see as her flaws or focus on our dreams for her future. We worry that our hopes might be dashed. But if you focus on an idealised future, the present is likely to disappoint; and if you let this show, the relationship that should fill your daughter with confidence will only fill her with self-doubt.

While some girls are as gentle, sensitive and helpful as we are told to expect them to be, we must never forget that others will be rumbustious, energetic, sometimes clumsy and filled with an urge to explore, play and tumble about. We should never see these girls as 'difficult'. Equally, we should never 'put upon' our compliant daughters, expect more from them than we would from our sons, or exploit any mother–daughter intimacy that may interfere with a girl's need to become separate and independent.

Parents

- imagine that your daughter has dropped all the characteristics that irritate you: she tidies her room, hangs up her coat, volunteers to wash up, never forgets anything; consider whether you are left with the same girl, and whether her essential personality has gone too

- list all of her pluses and minuses; balance each negative characteristic with a positive one; then add more pluses than minuses to the list

- let her live in the present, not with your fears for the future; she has many years ahead in which to grow and mature before she's an adult

- be aware that if she becomes who you want her to be, she will have lost herself

Teachers

- if a student becomes a 'clone', modelling herself on you or an ideal, she is likely to find it hard to take risks and handle making mistakes

- to help younger children appreciate who they are, outline them while lying on paper on the floor, then invite each child to fill in her shape with her characteristics

- ask older children to list 20 things they like to do, beside which they should list: the date when they last did them, a £ sign beside things that cost more than £3, an 'F' if they prefer to do it with a friend, an 'A' if alone, a 'P' if it needs planning, and an 'M/D' if a parent used to do it as a child; then they can then tell a story about their likes and interests

38 Don't make approval conditional on good behaviour

One of the things prospective employers are wary of when they interview people for jobs is any applicant who curries favour, seeks approval, avoids disagreement and seems not to have faith in their own judgement. Anyone showing these tendencies is rejected, for insecurity and uncertainty are unhelpful in the workplace.

Of course, we all have times when we feel unsure. However, some people suffer from insecurity more than others, and some are hampered by it most of the time.

The tendency often starts in childhood. Girls grow strong inside when they feel approved of, loved and accepted for who they are. If an adult's approval is conditional, and only forthcoming when a girl is being 'good' or has done well, she will be forever looking over her shoulder, manipulating her behaviour and creating distance between her instincts and her actions. Always having to play to the parental gallery, she will soon lose sight of herself and be unable to develop a sense of personal integrity or confidence in herself.

Parents

- accept that your daughter won't be perfect and that mistakes are not only inevitable but also important for learning
- see the funny side of her errors
- behaviour talks: she's not bad, just trying to say something; look behind any naughty behaviour for possible reasons
- disapprove of what she does, not who she is
- with older girls, you can disagree with what they want to do, yet still support their right to do it

Teachers

- be aware that reward systems for work and behaviour might lead unsuccessful girls to feel inadequate or disapproved of
- show approval towards all students: respect, show interest in and talk to each one, not just the co-operative and successful ones
- involve all students in decision-making, to develop their independence and self-esteem, and to demonstrate that you trust and approve of their ability to make realistic judgements
- value a wide range of skills

39 Hear her complaints

When my mum picked me up from my first school, she used to stand talking with her friends for ages. I was tired, wanted her because I'd missed her, and wanted to be home. Tugging at her didn't work, so one day I told her this. She said she hadn't realised, and changed straight away.

A very young child takes life as it comes, largely because she knows no different and is in no position to pass judgement. But as her sense of self and her speech develop, she begins to reflect and see the world from her own perspective. She becomes aware of her own desires and wishes, forms her own judgements, and starts seeing that things can, indeed, be different. When she puts this into words, she is expressing and risking her total experience of herself.

This is why, as soon as she is able to voice, or display, disappointment or dissatisfaction, her complaints should be taken seriously and responded to respectfully. Her self-confidence and happiness depend on it.

Parents

- let your daughter know that it's all right complain; pin a sheet of paper in her bedroom for written comments if she finds it difficult to confront you
- when you hear her complaint, it could be the first step towards compromise and an important lesson in conflict resolution
- hear her through: try not to be defensive or competitive if she complains
- be ready to apologise if she says you have gone too far

Teachers

- turn any complaint into a question or statement: 'It sounds as though you think this mark isn't fair because you tried really hard this time, is that right?', 'I think I need to explain myself more fully. Thank you for letting me know'
- if you are able to enter a student's world, see how things are for her and accept her perspective, you will be modelling respect, empathy and emotional literacy

40 Acknowledge her disappointments

Disappointments are part of growing up. Girls have to learn that they can't always get their way, and that when this happens, life doesn't fall apart. We have to learn to compromise, and sometimes to do without.

If parents set out to make sure that their daughter never experiences disappointment, they may end up enslaved. She will not learn to live with or overcome setbacks, and it will not help her to understand herself, because she will never have to decide which of a range of alternatives is really important to her.

However, there are times when disappointments should not be glossed over, when they should be not only acknowledged but also actively avoided. If the people your daughter needs to rely on and trust let her down frequently, it can lead to a profound sadness that may later develop into a mental-health problem.

Parents

- anticipate potential disappointments
- don't ignore or dismiss your daughter's sadness
- talk about it; show insight and understanding by letting her know you know, and say something like: 'I know you'll be disappointed, but we can't go bowling until next week' or 'I know you were looking forward to that a lot. I'm sorry that it can't happen and that I raised your hopes'
- remember that if she feels let down by you, rather than by not being allowed to have or do things, and she feels that you let her down frequently, she could begin to distance herself from you

Teachers

- most students will feel disappointed if they get a poor result, especially if they tried hard, though they may cover it up well
- try to acknowledge this when you give feedback, and if you think a pupil has made a special effort, acknowledge this too; tell her not to be down-hearted and assert your confidence in her ability to do better next time
- give her hope; discuss with her what she thinks needs to be done and what she can do differently, and end the conversation with a summary of steps she can take to raise her performance

41 Hold on to your authority

I thought I'd lost it. She wouldn't do a thing I asked. I felt completely useless, and became scared to have another go in case I got ignored again. Then I realised that, simply as her mum, I had all the authority I needed and I didn't need to prove it through shouting and screaming. I calmed down, thought ahead, decided on a few things I wanted her to do and stayed firm and fair. It worked, and we both feel so much better for it.

Authority can be demonstrated in many quiet ways. Adults have to find the right balance, so that they remain in charge, but don't get caught up in power battles to prove it. Taking firm decisions about your family's or class's routine – what you do when, or what behaviour is right for your family – is one way. Being unflappable, and showing trust that girls will co-operate as asked or expected is another. When you take clear responsibility for deciding things, you demonstrate your authority.

Parents

- all parents possess the authority that is vested in their position as a parent: you may have lost touch with your authority, but you can never 'lose' it
- be aware that things like threats and bribes, or shouting and shaming, are tools of power that children resent deeply; they will ultimately undermine your authority, not boost it
- when you assume responsibility for things, you automatically acquire and display authority
- you are the adult, so take the responsibility; if your relationship with your daughter has been going badly, tell her she hasn't been herself, and be constant about guidelines you set

Teachers

- if you trust a girl to behave as expected, and she in turn trusts you, this demonstrates your authority, and assumes a joint responsibility for resolving the problem
- if you make it clear from the start that your professional objective is every child's best interest, and you are able to convince students of this, when you make a mistake they will not lose faith in you, or themselves

42 Use reasons to explain, not persuade

I'm not sure why, but the more reasons I give, the more my daughter sits tight and refuses to do it.

Girls deserve to be given reasons: it shows respect for their right to know and their ability to understand. Hearing reasons teaches them how to put forward an argument and to be rational. But it goes wrong so often, because offering more than two reasons changes what you say from an authoritative command into a much weaker exercise in persuasion.

First, children frequently switch off at the first sound of a wheedling, pleading voice; they know what's coming and they feel manipulated. Second, we tend towards overkill by giving too many reasons. Children quickly learn that multiple reasons are a device to get them to agree, so they argue and refuse.

No more than two reasons are needed. Try saying: 'This is what I want you to do, this is why; now go and do it.'

Parents

- give no more than two reasons to explain why you want something done
- look your daughter in the eye as you tell her, so that she can tell you are serious
- then turn away, because this conveys the clear message that you expect her to comply; hovering suggests that you think she won't do it, and that she will need policing
- to avoid appearing too controlling when you want to say no, ask her to first guess and then tell you what your answer and reasons are likely to be

Teachers

- if a student asks for something outside the accepted rules, ask her to repeat the rule and anticipate what your answer is going to be; she then arrives at the answer 'no' without you coming across as negative
- give no more than two reasons to explain why a student has to do something

43 Be fun, fair and flexible

My granddad grew up in a mining village where all the families had very strict rules. But for one day, once a year, all those rules were dropped. The children were allowed to do what they wanted: knock on elderly neighbours' doors and generally go wild. They let off steam and everyone had fun. It sounded great!

While many girls seem to enjoy the safety and security offered by structure, none will thrive if she feels she is in the grip of an unrelenting disciplinarian.

Girls will give their respect and full co-operation only if the rules are clear but remain in the background, and if their daily experience is characterised by fun, fairness and enough flexibility for them to feel they are listened to, loved and treated as individuals.

Parents

- make time for family fun, outings and games with your daughter
- hiding birthday and other annual presents or chocolates around the house as a treasure hunt adds to the fun, and shows you have made an effort to do something special for her
- all children love family rituals, which can happen weekly, monthly or annually; if these can be fun, involving some relaxation of rules, such as a scheduled bedtime, they will be enjoyed even more
- flexibility, backed by a reason and agreed with a twinkle in your eye, won't lead to any loss of authority; insensitive rigidity will

Teachers

- curriculum planning and targets offer little scope for flexibility, but fun can be introduced into lessons through educational quizzes and games
- try to see mild pranks played on you as girls having fun and letting off steam; if you let these annoy or upset you, they'll target you again
- be creative: adapt a lesson to address an issue currently making the headlines or make it relevant to your students
- being fair means not just treating all pupils equally but also being sensitive to the reasons behind their behaviour

44 Rules reduce conflict

Discipline is the aspect of parenting that causes most parents the most heartache. It is also the thing most parents feel they do not get right. This isn't surprising, since there is rarely a 'right' answer: the rules have to shift as children grow and circumstances change.

Creating family rules is difficult because it entails balancing different people's needs and demands, managing different and developing personalities and sometimes compromising between different cultures and values.

Family rules that are mutually agreed and clearly understood undoubtedly reduce conflict. It's conflict that does the harm, not the rules. Clear expectations and well-established, consistent, daily patterns for you and your girls reduce the number of challenges. When she sees that you mean business, your daughter will stop testing you.

Parents

- conflict isn't avoided in the long term by giving in; your daughter will only learn that the more she pushes, the more she will get her own way
- all children like the security that rules provide
- being rule-abiding at home will help girls to be law-abiding later
- if you feel you may be losing control, prioritise; stand firm on the issues you care most about and drop the rest

Teachers

- clearly stated rules applied fairly and consistently throughout the school by all staff help children to know where they stand and feel secure
- involve the girls themselves wherever possible in agreeing the rules so that they don't feel dictated to and learn to take responsibility
- have explicit, graded consequences for well-defined breaches of the rules

45 Avoid wielding the tools of power

As a teenager, my mother was terrified I would get pregnant. She told me that if it ever happened, I would be out in the gutter. I was really hurt. How could she ever do that to me? After a huge row one day, I considered the best way to get back at her. I very nearly went out to have unprotected sex just to spite her.

The tools of power that adults are inclined to use are hitting, hurting, damaging children's belongings, bribery, ridicule, threats, sarcasm, shouting, emotional withdrawal and withholding food and liberty. It may be tempting to use these sometimes, especially when your patience is thin, but it will be counterproductive. Your daughter will simply find a way to get her own back, to preserve her self-respect.

Children deserve the best from their parents, not the worst.

Parents

- it is best not to force an issue when either you or your child is tired; let it go, in case it blows up in both your faces
- try using the 'soft no'; if your daughter does not respond the first time you ask her to do something, instead of raising your voice and issuing threats, repeat your request more quietly, making sure that you and she are looking directly at each other
- try trusting her to comply, giving one or two reasons, or using creative ways to get compliance

Teachers

- responding with instant punishments in an apparently arbitrary way is an abuse of power; be measured, fair and consistent to avoid resentment and maintain students' co-operation
- avoid using sarcasm and ridicule in the classroom; these are not appropriate tools for any professional
- don't take challenges personally; they may not be intended as such, and you run the risk of a communication breakdown
- people usually shout and throw things when their patience and skills are exhausted; consider team-teaching to refresh your skills if you lose control more than occasionally

46 As the adult, it's your job to repair

My daughter and I couldn't see eye-to-eye. We had one patch when we didn't speak for nine months.

When your relationship with your daughter breaks, however much you think she is at fault, it's your job as the adult to mend it. You have the greater wisdom and maturity, and the confidence and skill to achieve this. Refusing to acknowledge or communicate with your daughter is no way to teach her how to repair a relationship or to give her confidence in herself.

Parents

- when your daughter's behaviour is awful, try not to take it personally; she often will be acting like that to get attention or to protect herself, not to 'get at' you, unless she feels she has good cause; it is unnecessary to retaliate

- you don't have to win every battle: take responsibility for the bad patch and make the first overture towards peace

- make the second, third and fourth moves; trust is thin after a bad patch, so don't expect any immediate change; conciliation should not depend on immediate reciprocation

Teachers

- personality clashes between students and teachers happen; if one seems to rub you the wrong way, the onus is on you to sort it out; be open with the student, reflect on any past personal experiences that may explain your reactions, and take responsibility; arranging for her to work with a different teacher may be the only way out

- be aware that personal and professional stress can undermine skills and tolerance; if stress is affecting you, be open about your state of mind, apologise, state your needs clearly and your class will almost certainly offer co-operation and understanding

47 Discipline without dictatorship

All girls need boundaries and a framework of rules to help them control and manage their behaviour and fit into their family or school. Boundaries and rules help keep girls safe and acceptable to others; they also allow adults to show that they care about what happens to them and also to become involved in their world.

Reasonable rules and self-esteem go hand-in-hand. Clear boundaries make a child's world structured, planned, predictable and safe; they give it a rhythm and pattern. But this is achieved only when family discipline acknowledges and respects the needs and rights of the child. Discipline which is inflexible and which strives to humiliate will, steadily and inevitably, chip away at a girl's self-esteem.

Parents

- be clear – prioritise, don't have too many rules, and keep them simple
- be firm – but also friendly and loving; stick to your rules 90 per cent of the time, but be flexible when it really matters to your daughter
- be fair – because this is the best way to stop her becoming resentful
- be consistent – try to respond in the same way each time and get your partner to do the same
- keep your love constant – don't blow hot and cold
- set a good example – behave as you expect your daughter to behave

Teachers

- dictatorship no longer works in the classroom, if it ever did; military-style orders and insults are unacceptable
- as with parents, be clear, be firm, be fair, be consistent, be fun and be flexible, giving reasons whenever you adapt or relax the rules
- plan for variety and use different teaching styles to engage all students naturally
- ban threats, sarcasm, insults, ridicule and shouting from your repertoire

48 Punishment without humiliation

My dad used to make me sit at tea with no clothes on when I'd been 'bad'. And he thought he was justified because he hadn't laid a finger on me. I've hated him ever since.

There are effective and ineffective ways to show girls how to manage their behaviour. The use of punishments that humiliate is ineffective in the long term because they make children resentful. Though they may do as they are told at the time, their obedience is the result of force; this won't lead to future co-operation.

For children to learn effective lessons from punishment, it should focus on what they have done wrong, and not cause resentment, anger, bitterness or other bad feelings. In other words, it should be fair and leave a child's self-respect intact. If you put her down, there will be repercussions. Being frequently subjected to humiliating punishments will eventually cause a child to feel shame, guilt, self-doubt and ultimately self-hate, and it may lead to anger, hostility, destructive and self-destructive behaviour.

Parents

- if you use punishments, try to be clear, fair, consistent and sensitive to their effect; keep them brief and show your daughter that you still love her soon afterwards
- deal with only one behaviour at a time – don't pile on the complaints or punishments
- punish the act and not the person
- alternatives to smacking include: withdrawal of privileges; restricted use of a favourite toy or pastime; withdrawal of pocket money; being sent to a cooling-off place; a verbal telling-off, and an early bedtime

Teachers

- always give due warning of any punishment you might give
- make sure your behaviour policy contains clear and graded consequences for clearly defined behaviour
- if students challenge you and feign indifference, don't up the ante and impose harsher punishments
- ensure the punishment fits, and is relevant to, the crime
- avoid taking challenges personally, for when that happens the punishment may become personal too

49 Model effective conflict resolution

Constant conflict ruins relationships and tears families apart. Children are deeply scarred by conflict, particularly when it becomes physical and takes place between their parents. Family conflict lies behind much teenage despair that is expressed through depression and self-destructive behaviour.

But we can't get rid of all conflict. Different people inevitably have different and conflicting views and interests that have to be reconciled. Also, each of us has times when we feel exposed and vulnerable, when we are inclined to see comments and actions as challenges, take these personally and react aggressively, even where no challenge was intended. What we can do is to understand why and when this happens, and learn how to manage and resolve it when it does, so that it does not get out of hand.

Resolving conflict safely and satisfactorily takes emotional maturity and social skill. Children have to be taught successful attitudes and approaches by the adults who care for them.

Parents

- good communication skills are the basis for successful conflict management; listen to your daughter's case, present yours using only 'I' statements, not provocative 'You' statements, and seek a compromise
- acknowledge the feelings as well as the different interests that underlie the dispute
- you don't need to win every battle; sidestep the issue if it isn't critical, and avoid arguments when either you or your daughter is tired
- friction between siblings teaches about conflict; if older ones can be left together safely, let them sort out their problem in their own time; younger ones may need help

Teachers

- all schools should adopt a clear policy of non-violence, as set out in *Towards a Non-violent Society: checkpoints for schools*
- a table can be set aside for children who have begun to argue to discuss their differences safely
- all school staff, including mealtime and playground supervisors, can be briefed about non-violent ways to resolve conflicts
- personal and social education programmes should contain lessons on violence prevention and effective strategies to manage and resolve conflict
- how you respond to challenges in class sets a strong example

CHAPTER 6

Responding Sensitively To Setbacks

Life presents everyone, including children, with experiences that can knock them for six. It is part of every school's and carer's job to prepare the girls they look after for these distressing times. Being bullied or excluded from groups, families breaking up, friendships ending, bereavement or separations, and disappointments at school or on the sports field are common experiences. Children can't be protected from every possible hurt, however much parents may wish to and however desirable this is – and it may not be. But how can adults help to rebuild a girl's self-esteem when all she wants to do is run away and hide?

Most of us will know a child who seems to be made of rubber and wears a permanent grin. Nothing seems to get her down. Whatever the problem, she has the ability to take the knocks, keep her attention fixed on a better future and make her way steadily, knowingly and confidently towards it. Such a girl is called resilient. She lives with and through problems, gets herself back on course, can move on and even use setbacks to strengthen and enrich herself.

Recovery from difficulties is more likely when adults respond sensitively to a girl's setbacks. When they manage her confusion and dejection well, she retains self-respect and self-esteem to face the world again and treat the obstacle as an opportunity to grow and learn.

Resilience is sometimes described as the ability to bounce back; however, this misrepresents what happens. Staying power involves action, not reaction. People with stamina and persistence think, feel, perceive and understand themselves and their situation in a way that enables them to remain positive, active and able to recognise the lessons to be learnt. Resilient girls overcome problems and recover from obstacles by using a variety of skills, attributes and strategies, often helped by adults.

What gives girls staying power? Recent studies of children who recover from setbacks show that personality plays a small part, but more helpful factors are a tendency to see themselves in a good light, having at least one good, close relationship with an adult and not being exposed to too much difficulty.

Girls who bounce back have a good sense of their own worth and abilities, believe they can shape what happens to them, are good at solving problems, mix happily with others, maintain friendships and are generally optimistic about life. Parents and teachers can help girls by encouraging this and by responding sensitively to them.

50 Give her safe time to talk

After my parents split up, I went to my dad every other weekend. The only time he asked me how I was was during the short journey home in the car. He wasn't really interested. He was making it safe for him to talk, not for me.

When girls feel close to someone and are able to make sense of whatever has happened to them, they are more likely to recover from setbacks. Talking usually helps, but the time and place to talk has got to be right. There's no point in raising the issue in a place where the conversation will be overheard, where either of you feel uncomfortable, or when there's insufficient time to explore and come to some conclusion, even if this is 'We need more time to talk'.

This is particularly true if your daughter is ashamed about the event or if she feels responsible – if she has been bullied at school or been found cheating or stealing, for example. Children sometimes feel guilty about problems for which their parents are responsible, such as the breakdown of a marriage. Even if they don't truly believe they're to blame, they may still feel that they should have prevented what happened.

Parents

- avoid being intrusive; before you talk, check that your daughter's ready
- make quite sure that there is enough time to talk before you begin
- make it clear from the start that the conversation is confidential
- if you want to pass anything on, get her permission: 'Do you mind if I tell Dad/Mother/your teacher what you've said?'
- if she finds it hard to talk, suggest going for a walk together; tell her you are ready to talk when she is, but let her take the initiative
- girls sometimes prefer to work things out in their own heads rather than talk; be aware this can be equally successful

Teachers

- befrienders schemes are useful in the upper primary and lower secondary school years, and allow pupils to talk to their peers
- in the early secondary years, form teachers play a vital role – they should be selected for their proven abilities in dealing sensitively with pupils
- any references to personal issues or problems in pupils' work should be taken seriously and responded to sensitively, not ignored

51 Fortify her heart, don't thicken her skin

Many parents believe the best way to arm their daughters against verbal attack and disappointment is to thicken their skin, to get her to 'toughen up'. They achieve this in two ways. Some dole out lots of hurtful words and crush expectations just to get a girl used to it, to create an immunity to pain. Others constantly tell their daughter that she should not be feeling the way she does; that her emotions make her vulnerable to the harshness of life and to other people. They tell her to deny or repress her feelings.

A far better way to protect her is to strengthen her inside – her heart and her self-belief – not her outer shell. The other route not only does great damage to a girl's self-esteem and self-understanding, in the process, it also cuts her off from her essential self. Suits of emotional armour stop feelings coming out as well as going in, so they offer no long-term help with managing relationships.

Parents

- build up your daughter's inner strength: trust her; see her as competent; let her have some autonomy over her life; tell her you love her; and respect her view of the world
- if she's in trouble, understand her feelings, support her and help her through; don't put her down by saying: 'Why did you let yourself get into this mess?'
- encourage her to follow and trust her instincts: 'encourage' means to give her courage
- give her assertive words and phrases to use when she feels threatened, such as: 'I don't know why you need to say/do these things that hurt others' and 'I don't have to listen to you' and turn away; discuss which ones she'll be happy to use

Teachers

- constantly monitor the different ways in which you talk to girls and boys
- don't fall victim to expressing stereotypical views about either girls or boys

52 Nurture her self-respect

When things go wrong for your daughter, she is quite likely to feel embarrassed or even humiliated. She may think that she has let either you or herself down, which may make her feel very anxious if you are generally intolerant of her making mistake. It is natural and quite common for children to keep problems such as being bullied to themselves. Especially if and when her problem becomes public knowledge, your daughter may be ashamed of the position in which she finds herself. Her self-respect will have been damaged, even if she is not really to blame.

In order to pick herself up and face the world again, what she needs most is a double dose of self-belief and self-respect. However you feel about what has happened, your prime role must be to help repair the damage done and to leave your daughter with enough self-respect to begin again and have another go. Humiliating her won't work.

Parents

- your daughter will learn to respect herself if she sees that you respect her
- those who are frequently blamed and shamed will find it hard to hold on to their self-respect; but they can retain it, and make good their mistakes, if they are guided sensitively, without criticism or insult
- self-respect grows in part from the experience of being given responsibilities and carrying them out successfully
- children shirk responsibility and will not learn self-respect if they are allowed to ignore any painful consequences of their choices or behaviour

Teachers

- show respect for your students, so that they can respect themselves
- help them to see the good in themselves and in what they achieve
- independent learning helps pupils learn to trust and respect their own judgement and not to hold back due to fear of failure
- encourage pupils to respect each other – to listen respectfully and share successes; in small groups, ask them to share a success, accomplishment or achievement they had between certain ages – before they were 10, between 10 and 12, and so on, as is appropriate to the group

53 Avoid shame and guilt

Shame and guilt are natural human feelings. When people have specific duties and responsibilities, and they fall short of meeting them, it is natural that they feel some shame or guilt. So why is it so important for a girl's self-esteem to protect her from too much shame?

Shame and guilt are difficult and destructive emotions for all children. They create doubt and confusion: the child has let herself or someone else down, and probably doesn't understand why or how. Just sometimes, it is appropriate for a girl to feel shame or guilt: for example, if she's old enough to be clear about what she should be doing, can manage her behaviour and be held fully responsible for it – if these feelings come from within herself. Punishing your daughter by making her feel ashamed and guilty won't work; instead of changing her ways, she is more likely to suppress the uncomfortable guilt feelings and deny any responsibility for the problem.

Parents

- make it clear to your daughter that any relationship or marital breakdown you are suffering is not her fault
- even if she has tried to sabotage a new relationship of yours, focus on what worries her about the change rather than blaming her
- concentrate on addressing the practical consequences of her difficult behaviour; your shame or embarrassment are your problem, not hers
- if she is being bullied, talk about what she might do to prevent it, and inform her school; don't tell her she is weak
- it's for her to decide how she feels about her mistakes, and for you to decide the consequences

Teachers

- instead of rubbing a girl's nose in her errors, make your expectations clear
- it's all right for a student to feel guilt or shame spontaneously; it is not all right for you to make her feel guilty or ashamed, however
- sarcasm, ridicule and teasing rarely spur children to greater effort, for the shame they engender undermines confidence
- don't punish the whole class for something only a few have done; it arouses widespread resentment, not repentance in the guilty ones
- don't blame students for something for which you are responsible

54 Offer extra closeness

When a girl experiences stress and difficulty, she will need the reassurance of the adults she depends upon even more. Yet if parents feel any anger or disappointment, they are more likely to withdraw than to draw closer. Children understand themselves through their key relationships – those they have with friends, family and caring professionals. If these supports disappear in her time of need, a girl will doubt herself even more profoundly.

Not only will she lose confidence in herself and question her identity; she will be further confused and unsettled by the altered behaviour of those she relies upon.

All changes are stressful for children. In times of change and difficulty, try to stay close to your daughter, and be around more.

Parents

- changes that may upset your daughter can include: moving house; starting a new school; illness in the family; bereavement; a new relationship or yours, or difficulties in your current one
- depending on your daughter's age, you can be close to her by: staying with her while she bathes; sitting on her bed at night; sitting next to her while she watches TV; keeping her company as she walks to the bus stop; giving her lifts in the car; and chatting as you drive

Teachers

- spend time with younger girls in the quiet corner
- ask pupils to help you with tasks
- find time for little personal conversations
- constantly offer yourself as someone with whom a student may 'talk things through' – or suggest other people who could do the same (friends, parents, form teacher or befriender, for example)

55 Help her see the lessons to be learnt

Don't despair; start to repair! Parents and teachers offer effective help when they remain non-judgemental and let a girl work out for herself what went wrong or what she did wrong, and how she can move forward again in a practical and confident way. There is always something to be learnt from setbacks. Rather than playing the victim and blaming someone else, your daughter has the option to come through difficult experiences stronger, wiser and more competent.

Stay positive. Instead of asking, 'What will you avoid doing next time?', you can ask, 'What might you do differently next time?'

Not all setbacks will quite fit this model. If a child is feeling down because she has been bullied, had a quarrel in the playground, or been dumped by a boyfriend, she will not necessarily have done anything wrong. But there will still be conclusions to draw and the solution should come from her. Problems are best reflected and acted upon, not recoiled from or smoothed away by an over-protective parent.

Parents

- unravel and break down the problem; if something went wrong, there will be practical reasons why
- don't let your daughter tell herself that she's 'useless' or that she'll 'never be any good'; setbacks should not be seen as unalterable omens of her future, but as learning opportunities
- ask her what she thinks went wrong; don't tell her what you think unless she asks or says she's happy to listen
- identifying the lessons to be learnt is energising; when she can see what went wrong, she will know how to put it right and regain control

Teachers

- making a student aware of the control she has over her daily life enhances her self-concept; ensure that organisational devices such as planners and homework diaries are used regularly and effectively
- we find out about ourselves through taking responsibility; coping with the consequences of our mistakes provides such an opportunity
- encourage students to develop the habit of asking the questions: 'What are the lessons to be learnt?' and 'What have I learnt?'; they both foster reflection
- if a pupil fails to do her homework, ask for a realistic plan for making up the lost ground

56 Hear her side of the story

When a girl has a setback, depending on the nature of the difficulty, it is common for parents either to take it personally and feel shamed by it, or to see the problem as a non-issue. This can encourage them to belittle their daughter's fears and concerns. If you have heard only the bare facts of the case, do not assume that your daughter is at fault, or tell her off for putting you in a difficult position with, for example, neighbours, the police or the school.

If she is upset about something you consider minor, try to see the issue from her point of view. Adults and children perceive things differently, so avoid applying your judgements unthinkingly. Saying things like: 'Stop making such a fuss', 'It will pass' and 'I can't understand why you're bothered by that' won't help your daughter resolve her problem; encouraging her to think it through will.

Hearing her side of the story shows that you respect her and want to treat her fairly and seriously. It will help to maintain her self-esteem.

Parents

- let your daughter tell you about her disappointments, and don't belittle them
- assume the best of her, not the worst
- in arguments between brothers and sisters, hear each one out, then ask them in turn what they see as a possible solution
- a child doesn't always want you to solve her problems; she may just want to sound off and be listened to; keep your opinions to yourself

Teachers

- don't assume a pupil is in the wrong if she has a grievance against a teacher; suggest she goes to the pastoral or year head to state her case, somewhere neutral where she will be listened to and taken seriously
- avoid making hasty prejudiced judgements; just because a girl has a reputation for being troublesome does not mean she is implicated in every problem; hear her side of the story

57 Identify the danger times and signs

When Debbie was 12, her parents separated and her world collapsed. She had a week off school and, when she returned, she had trouble keeping up with deadlines, maintaining standards and doing homework. She started to break down over little things. It took her about a year to catch up, and three years later she was still vulnerable to any kind of setback.

Children's resilience inevitably lessens when their parents or carers split up, and during bereavements and other forms of family upheaval. They are also particularly vulnerable when they start or change schools. Most parents appreciate that children may be unnerved when they enter infant or secondary school, but even moving from infant to an adjacent junior school can be unsettling.

Girls' confidence is fragile when they pass through key developmental stages, too – at the age of about eight and at the onset of puberty and adolescence. At these times, girls need plenty of attention.

Parents

- your daughter will need a lot of support and attention when she starts secondary school, and for the following two years
- situations that can destabilise children include:
  - domestic violence
  - racial abuse and harassment
  - constant criticism and abuse
  - multiple home/school moves
  - divorce and separation
  - bullying
  - family re-formation
  - gaining step-families
  - bereavements
  - temporary absence of a parent
  - illness/disability in the family

Teachers

- signs that children are in emotional trouble include being:
  - withdrawn, alone, lonely, sad, prone to tears, tired and inclined to fall asleep
  - unable to concentrate
  - reluctant to attempt work
  - very aloof and rejecting help
  - unable to follow classroom routines
  - needy or contriving helplessness in order to get constant reassurance
  - likely to destroy their own or other children's work
  - afraid to mix with others
  - aggressive, clingy or dependent
  - the class clown
  - a frequent target for bullying
  - a liar, thief or cheat
  - late often
  - absent frequently

58 Dealing sensitively with peer pressure

Peer pressure, along with bullying and drugs, is a subject that frightens many parents. Quite apart from the expense of providing the 'in' trainers, designer clothes, and the very latest in computerised home entertainment, we like to believe that our daughters will be sufficiently independent to withstand the pull of her peer group, especially when the group's activities are dangerous or illegal.

Children generally love to conform and hate to be different. Their earliest flirtation with independence from parents and self-expression is frequently via the safe route of fashion, and the younger they are when they make choices about clothes, hairstyles or how they spend their time, the more likely this is. Not all peer groups are insidious.

Parents

- don't drive your daughter into the arms of antisocial friends by being constantly critical; the best way to help her resist negative peer pressure is to nurture her self-esteem and give her inner strength
- if she wants expensive clothes or outings, ask that she contributes something towards it, and make this possible with regular pocket money, or part-time earnings if she's old enough
- ask her for her definition of a friend, and question whether people who won't allow her to be different are truly friends
- remember, girls who are vulnerable to peer pressure are impressionable; help your daughter feel acceptable as she is

Teachers

- be aware of the power of peer groups to divert some girls from their studies; a 'befrienders' scheme can offer vulnerable girls a listening ear
- discuss the issue of peer pressure in assembly, English or PSE lessons
- many girls who are seduced away from learning are vulnerable because they are already failing; identify those who may be at risk as early as possible, and offer them mentoring in order to keep them on track

59 Keep her informed of developments

One of the most important needs we all have is to be kept informed about changes and events that affect us. We get very annoyed if employers, partners, the local council or neighbours do things without letting us know in advance. Children need to be told things too, especially if their families are experiencing changes.

If you and your partner are having a trial or permanent separation or going through a divorce; if someone in the family is ill and needs a lot of medication, or has to go into hospital, perhaps for an operation; if you are looking for a new school for her, your daughter will want to know how things stand. All children can be panicked into thinking the worst, or disappointed by an unrealistic dream, if they have only sketchy information to go on.

As stated earlier, if you keep your daughter informed, you show respect for her right to know, for her need to understand what's happening to her, and for her ability to take in that information and use it sensibly.

Parents

- children should be kept informed about: events and changes (before, during and after they happen); feelings (yours and theirs); decisions and facts
- anyone who feels uncomfortable about discussing issues such as going into hospital, dying, divorce or moving house may prefer to raise the matter using a children's book that explores the subject through a combination of story and factual information
- children need to make sense of their world and what is happening to them; keeping them informed of developments is an important way to do this

Teachers

- ensure that pupils know exactly why something is happening, from the very basic, such as what the aims of a particular lesson are, to why a pupil is being moved to a different group or who'll be taking the class while you are away, etc.

60 Find good support groups

When my mum died, my life fell apart. My friends at school and the swimming club were the two things that kept me going. The club got me out of the house and made me feel normal. My friends were just great.

Research into what helps children and young people to manage difficult personal events shows that they do better when they are involved in a range of groups and feel part of a wider community.

Belonging to a group can help girls feel secure about who they are at a time when they might question their identity. It is reassuring because their commitments maintain normal daily and weekly patterns and routines. It also enables them to receive the understanding and support of people who know them and have time for them.

Parents

- help your daughter attend her regular groups and clubs during troubled times
- if she has few commitments outside school, see if there are other activities locally that she can become involved in
- if she's having problems at school, out-of-school groups can help her start new relationships with a clean sheet
- if you think she's mixing with the wrong set of friends, keep an eye on her
- don't allow your need for her company to isolate her from her friends

Teachers

- encourage vulnerable girls to sign up for lunch-time clubs and after-school activities where they can develop friendships and new skills and rebuild their confidence
- vulnerable girls may benefit from being kept together in stable class groups, which offer security and support

61 Minimise conflicts at home

This is a hard one, for arguments are a natural part of family life. However, it is very distressing for girls when home is, in their eyes, characterised by conflict, and it is clear that domestic violence and conflict are associated with girls who have particularly low self-esteem.

Conflict is unsettling. When girls are going through a difficult time, and already feel unsettled, parents should take extra trouble to avoid arguments, both between themselves and with the girl concerned. When parents fight, verbally or physically, their daughters can feel under pressure to take sides, which splits them down the middle.

How can parents know when their family conflict is greater than 'normal'? Features that probably make a difference include: frequency (obviously); who is involved; what the disagreement is about; whether it becomes noisy or personalised; whether verbal or physical violence, abuse or bullying are involved; and whether and how the conflict is resolved.

Parents

- consider banning arguments during difficult times
- encourage everyone in the family to write down their complaints instead; these can be discussed at a set time; say, once a week
- remember that personal insults, shouting and abuse will damage your daughter's self-esteem
- discussion and debate are healthy, and are not the same as conflict
- teaching and modelling ways to resolve conflict will be of life-long benefit to your daughter

Teachers

- conflict at home can affect girls' school work, especially when it is serious enough to be described as domestic violence, so it needs to be treated seriously by the school
- discussions about conflict at home and at school, why it happens and different ways to resolve it, should be part of all pupils' personal and social education programme
- if girls hear that others experience the same problems, it may help them to cope

CHAPTER 7

Supporting Her Learning and Personal Growth

Children do well when their parents are openly interested in what they do and support them, and when they have a strong, realistic sense of themselves and what they can achieve. 'I believe I can do it' is as important as 'I want to do it', and both are far more important for long-term achievement than 'I was made to do it.'

Given that the key to successful learning and personal development is self-motivation, what can parents and teachers do to encourage this? From research undertaken in industry, we know that the best motivators are people who: try to see merit in an idea even when it is different from their own; accept mistakes if lessons are learnt; are easy to talk to, even when under pressure; have consistent high yet unrealistic expectations; encourage people to develop themselves; and give credit when credit is due. From sports psychology, we know that the best coaches focus on improving technique and skill, not on the target of winning, and discourage athletes from judging themselves by results; make rewards reflect achievements; teach individuals to manage their own mistakes, learning progress and reduce anxiety by finding out what is causing it and addressing that directly.

Learning and growing inevitably involve change and taking risks, the unpleasant possibility of outright failure, confronting personal limitations, and the excitement of potential discoveries. When girls feel competent and good about themselves, they are likely to be positive about their present and future skills and talents. Confident girls are happy to change.

Girls who feel very unsure of who they are will feel pessimistic about what they are capable of achieving. But they will often avoid acknowledging this, and sidestep failure and change by claiming they know it all already or have no need to know. Seeing learning as irrelevant to one's future is simply one further, highly convenient and notionally respectable way of avoiding taking responsibility for success and failure.

Girls who fail to master the basic skills at school may eventually stop trying to achieve anything. Instead, they may focus on what remains within their power to control and give: their body and their sexuality. Helping girls to grow in confidence and explore their practical and intellectual talents requires supporting their school work and giving them reason to consider themselves worthy of effort.

But when girls are depressed, anxious, angry, feel let down by adults or are preoccupied with any problem, they cannot concentrate on their work. During these times, try to maintain routines and to offer as much emotional support as possible.

62 Encourage and value a range of skills

As a young girl, I followed my father about the house when he was mending things. I loved to see how things worked and I learnt a great deal. When I set up my own home, I saved lots of money by doing some quite big jobs myself.

Every girl will have many talents. She may be good at football, dancing, drawing, constructing paper models, cooking, climbing trees, rollerblading or ice skating. She may be knowledgeable about insects, animals, dinosaurs, gardening or pop music. Or she may be best at thinking things through or organising herself and others. Perhaps she is quick to understand how someone else feels.

Children do best when they are well informed about or successful at something, and when they enjoy a range of activities. It is a parent's job to identify and value the various talents and skills their daughter possesses, and draw her attention to them. Academic prowess should never be the sole criterion used to evaluate a child.

Parents

- try to broaden the base of achievement: let your daughter sample a range of activities and skills from those that are available locally
- every child will benefit from believing she is good at something, such as doing jigsaws, playing with other children, being creative with colour and paints, or skilled with computers
- keep television-watching in check: your daughter needs balance and variety to get the best from herself
- try to involve her in the practical DIY and domestic tasks you do around your home

Teachers

- find something that each student is good at, tell each one what she does well, then work to develop her strengths and interests
- if a girl has an unusual talent, encourage the other children to respect her skill; but first discover it yourself!
- work hard to break down gender-stereotypical choices for options in Year 9
- set up after-school and break-time clubs to introduce students to new interests

63 Support and encourage, don't control and push

Support and encouragement will give your daughter the energy to go that extra mile when she's struggling. But support and encouragement may turn into controlling and pushing; while the former is good, the latter often backfires, leaving an exhausted and resentful child who may opt out of success.

The words themselves suggest when the line may be crossed. 'Support' implies sharing the strain or burden. 'Encourage' means to embolden – to help someone to be brave about doing something new or difficult. By contrast, adults who control and push girls don't share their burdens; they add to them. They imply that a girl cannot be trusted to manage something on her own. The expectations and targets they set are more likely to undermine her self-belief and sap her courage than to build it up.

Parents who control and push tend to finish tasks for their daughter; commit her to too many out-of-school activities; point out mistakes immediately; use threats and bribes; set new goals in quick succession; complain to the school often; hover, and get involved in her homework, even rubbing out mistakes she makes.

Parents

- take an interest in your daughter's hobbies and pastimes; watch her doing them and ask how things went, especially when she made a special effort
- offer to take her where she needs to go, discuss any problems she has and try to answer her questions
- listen to what she tells you and share her excitement about her ambitions and dreams
- trust her to set her own goals in a time-frame she can manage, and let her work using her own preferred learning style and work patterns
- remember, children are people, not puppets or performers; if you control, push and pull, their life can become an act

Teachers

- give each student detailed information about the progress she's made and what she still needs to do
- help her to devise a plan of action if she gets stuck, to keep her on course
- be enthusiastic about her improvements
- use stars and incentives carefully, because they can be manipulative; if a girl decides not to co-operate, she may be left with nothing to work for
- letters, postcards or certificates sent directly to a girl's home can allow parents to enthuse about recognised achievements

64 Trust that she'll manage

I sit with her every night when she does her homework, and then I check it. If it's no good, I make her do it all again.

A fundamental, unspoken and mutual trust is created between mother and infant at birth. At that moment, the infant must trust that her mother will care and provide for her, and the mother trusts, in turn, that the infant will love and need her. As your daughter grows older, if you do not trust her, it insults her. Growing up involves facing new situations and experiences at a rate most adults would find enormously stressful. It takes great courage to do this. Your daughter needs to believe in herself to cope with new things confidently. Imagine the hurt, and the devastating blow to her pride, if the person whose opinion she most trusts assumes that she will fail – often before she has really tried.

Parents

- trust your daughter's competence, her ability to see a task through, her judgement, responsibility, and her capacity to take and trust and praise her ideas

- when she's attempting to do something, tell her you believe she can do it; walk away, don't hover over her expecting problems, or say 'Can I help?'

- managing a part-time job and school work is possible with careful time management; let her try it before you ban it

- if she starts a new school, trust that she'll cope; saying, 'I hope you will/will not...' suggests that you fear the opposite!

- if she's finished revising, don't push her do just a bit more

Teachers

- ask a student if she thinks she can manage; if she says yes, affirm her self-belief

- occasionally, work presented as a challenge does get results; a girl will strive harder if told: 'Now I'm not sure you'll manage this'; however, this tactic must be used sparingly during adolescence, when girls' self-esteem is fragile; only use it when you firmly believe a student can deliver and is unlikely to view the challenge as a personal slight

65 Make it safe to make mistakes

Your daughter will need bucketfuls of self-belief to make the most of her potential, but she won't have this if you criticise her every time she makes a mistake.

Mistakes are an essential part of learning. 'Everyone gets scars on the way to the stars,' wrote the songwriter Fran Landesman. Mistakes are also useful, because they shed light on the task in hand. They show what does and does not work, and what needs to be done differently. Mistakes tell a story, and it is the story we need to understand.

Companies now understand this. The motto of one American IT company is 'If everything you do is a success, then you have failed', because mistakes show that someone has the confidence to take risks and be creative. Business today rewards staff for managing errors rather than punishing people for making them.

If you fear and deny your own mistakes, you won't help your daughter. Love her for who she is, mistakes and all, not only when she is perfect.

Parents

- be honest and light-hearted about your own mistakes, and point out what you have learnt
- give your daughter time to notice her own errors, or say, for example: 'I can see two problems here, can you?'
- careless homework could suggest she's not doing it in the best place, at the best time or in the best way; seek her views
- if she lets herself down under the pressure of exams, she might be investing too much in her results or worrying about your reaction to them
- unexpected mistakes could mean that she has misunderstood something, hasn't done enough work, hasn't used the right method or that she is worried

Teachers

- be honest about your mistakes, state what you've learnt, and apologise for them if relevant
- if a particular girl is more careless than usual, it could be because of your teaching method; alternatively, she may be distracted, distressed, or anxious about making errors
- raise the class's awareness of their attitude to making mistakes by encouraging discussion, and note whether the girls react differently from the boys
- unlike boys, girls (in general) prefer clearly defined tasks that do not entail risks; but risk-taking is something they need to get used to; give them open-ended, risk-focused work, and reflect as a class on their reactions to it

66 Have realistic expectations

My six-year-old daughter's report was positive about every aspect of her work and social development. When I said, 'Well done', she turned away and choked, 'it was a stupid report. What's the point of telling me I'm good when it's so easy?'

People perform according to expectation. They tend to live up to – or down to – their reputations. Far too many children fail for far too long simply because they are not asked to do any better. They never realise what they could achieve because they never stretch themselves, and they believe the limits of their capability to be those implied by the easy targets they set.

However, parents and teachers have had it so drummed into them that they must raise their expectations, we are now in danger of going too far the other way. A target set too high is as unhelpful as one set too low. Challenges must tempt, not intimidate, deter or exhaust. Targets that are too high can lead to failure, shake your daughter's confidence and make her believe that succeeding is the only way to gain your approval.

Parents

- ask your daughter what she thinks she can manage
- help her to devise clear plans for meeting her goal
- invite her to think ahead about any problems she might meet and how she'll manage any setbacks
- invite her to select short-term targets, which seem more achievable, as well as longer-term ones

Teachers

- make sure the target, time and quality of conditions for the work you set are clear, and let your students assess whether they have met the conditions
- encourage planning and reflective skills
- ask for a detailed study plan if you detect over-confidence – don't tell a student that her target is unmanageable
- ask pupils whether they're ready for the next challenge, or want time to consolidate their new achievements
- be clear, practical and realistic about your own targets

67 Read and learn together

Technology has simplified our lives so we no longer lead simple lives.
– Trond Waage, Norway's Children's Ombudsman

So close your laptop to free up your lap for your child.

Girls are known to develop speech and language ahead of boys, which can help them to read at a younger age. Despite this, all children need to grow up in their own way and at their own pace, and there is no value whatsoever in pushing a girl to read before she is ready to do so. Nevertheless, all children find learning to read easier if they are familiar with books, are interested in what they can learn from them, enjoy just looking at the pictures, and associate books with a cosy intimacy with people they're close to – male and female. Grandparents, parents, aunts and uncles, step-parents, boyfriends and girlfriends are all people who may have a special relationship with a girl, can make her feel special and loved, and help her become a fluent reader by sharing books with her.

Parents

- young girls are usually happier with stories than boys, who usually prefer factual books; non-fiction will enhance your daughter's general knowledge and attention to detail, and fiction her literacy and imagination
- keep bedtime reading for pleasure – she'll be far too tired to be put through her paces at the end of the day
- girls often read less in their teens, when friends and freedom beckon; try reading alongside her, with her music playing, to recreate the earlier cosy intimacy; if this encourages her, she may even decide to read a bit of her book aloud to you

Teachers

- choose a range of subjects for class reading that will suit boys and girls in turn, but recognise the benefits of fiction
- develop paired and shared reading programmes that train unconfident, though competent, readers to assist younger or weaker ones
- use book displays and other means to portray reading as a male and female activity
- poetry, especially humorous verse, is enjoyed by girls and boys alike, and can be a useful introduction to more formal literature

68 Show interest, but don't be intrusive

My father never once came to a parent–teacher meeting, to watch a sports event or see me in a play. He reckoned he'd done his bit by earning the money to pay the bills. It upset me a lot, and I ended up dropping out.

This girl had successful parents. Her mother was interested, but her father wasn't. She obviously wanted both parents to know about everything she did that was special and significant to her.

It doesn't take much to show an interest in what your daughter is doing: a question or two to initiate a conversation and follow up a point; a small moment out of the day to connect with her life and thoughts. Attending evening meetings, sporting events, etc., takes a little more effort, especially if it means leaving work early, but putting yourself out and being there build a feeling in a child that she's important to you and that you take her school and her learning seriously.

But adults can reach the point where they ask too many questions, often for the wrong reasons. Children can then clam up, seeing the questions as an intrusive inquisition.

Parents

- ask the right questions for the right reasons; ask something straight, or leave well alone; don't try to get information deviously, because your daughter will almost certainly figure out what you're up to, then shut you out

- wait for her to tell you her exam or test results; don't make them the first thing you ask about

- at the end of the day, instead of asking her 'What happened at school today?', you can tell her what you did, and then say, 'Is there anything you want to tell me about your day?'

- pointed questions about marks, getting into trouble, or what she did at break-time, speak volumes about the issues on your mind

Teachers

- for secondary school parents' evenings, don't rely on 'pupil post': use the official mail service or telephone to ensure that all parents have the opportunity to attend

- at parents' or new-intake evenings, and in school newsletters, constantly emphasise the important role that parents can play

- stress that a positive role model is vital

- find the time to take an interest in certain personal things that are important to your students, but don't persist if this makes any of them uncomfortable

69 Develop persistence: help her to see tasks through

My daughter is at university now, but she still phones me up in a panic when she's got an important essay to write, and exam time is even worse. She loses her confidence completely, and then she begins to question what life's all about.

Girls tend to be more persistent, dogged, and dutiful than boys. They seem to be able to defer rewards for longer, are less likely to be bored and more tolerant of unstimulating teaching. The move towards course work, projects and modules in national examinations suits girls well.

However, most girls have times when they doubt themselves. Their vulnerability is their self-belief rather than their stamina. Exams bring any lack of confidence to the surface. The best-laid plans can come to nothing without 'stickability'. Her motivation can be strong, the target can be clear, but if your daughter doesn't have the confidence to stay the course when she feels under pressure, the work she has already invested may be wasted.

Parent

- if your daughter gets stuck, empower her; don't belittle her or give her the answers, but help her find them herself, to give her the confidence to do it on her own next time; then back off

- at any sign of flagging from her, show interest and ask her to read or show you what she's done, or tell you what she has enjoyed and what's been hard

- check that she believes the target is achievable, and that she has a practical plan

- reinforce her self-belief in every aspect of her life whenever you can

Teachers

- all children find it easier to concentrate if work and any rewards are delivered in short, sharp chunks; give targets a high profile, and make the routes to each one clear, so that interest does not wane

- if a pupil gets stuck or loses interest, jointly prepare a clear plan to get her back on course

- make sure a girl knows why a piece of work is good, so she can't put it down to fluke and maintain her low opinion of herself

- don't constantly move the goal-posts further away; to gain real confidence, girls need to feel that they are on top of their work, not constantly struggling to attain a higher goal

70 Support the school

Although, by and large, girls today do well academically, they are growing up exposed both to a culture that questions the value of academic success and to an economic climate that offers instant fortunes and makes exams seem irrelevant. The temptation not to play the school's game is great, and not every girl will succeed academically. We can't change that culture, but we can model the importance of learning as a life-long process and accept that home and school need to support each other, in partnership.

If parents respect schools, and schools respect parents, there will be fewer cracks for girls to fall through, particularly during the earth-moving time of adolescence. Parents who distance themselves from their daughters' school not only create split loyalties, but make it easier for girls to team up with the tear-aways rather than the teachers.

Parents

- talk your daughter's school up, not down; avoid complaining about it in front of her, even if you're paying fees that are too high or having a political debate
- wherever possible, both parents should attend meetings about their daughter's progress in school; an absent parent can telephone the teacher for a personal summary
- try to be free to watch your daughter take part in school events; avoid saying: 'Not another thing that school/club wants me to do!' or *she* wants you to do
- attending fund-raising events with her will help her to feel part of the school
- help her to remember and collect the things she's been asked to bring to school

Teachers

- try to let parents know when things are going well, not only when there are problems; when parents feel proud of their daughters, they usually treat them better
- take parents' worries seriously, and respond to their concerns with respect
- avoid appearing to criticise a girl's parents: 'Didn't your mum know you'd need sandwiches for today's trip/check that you'd got everything?'
- make parents' evenings focus on what parents can actively do to help their girls set appropriate targets; stress the importance of leisure, too

71 Respect her teachers

It's far harder now than when I started teaching twenty-five years ago. It's not so much the constant changes to what we're expected to teach or the extra paperwork; it is the parents and the students, who show us so little respect now. When children hear parents doing us down at home, it's hard for them to accept our authority and take work seriously when they're here.

Children will not make the effort to listen and concentrate if they do not trust or respect their teacher. Constant carping about teachers at home, especially about a particular one, will encourage disrespect and challenging behaviour at schools.

Schools and parents need to work together as partners, respecting each other, not fighting or criticising one another.

Parents

- at all times, try to put across the teacher's perspective and be realistic about his or her commitments, even if you side with your daughter
- teachers are people too; they have personal lives and sometimes go through hard times; they like to hear good news as well as bad; most do their best and are stretched close to their limit
- don't be timid about telling a teacher what seems to work best for your daughter – they can't know everything
- it's only fair to the teacher and your daughter to tell the school if there's a problem at home that might affect her behaviour or work in school

Teachers

- we should all have to earn the respect we feel is our due; try to see things from the parents' angle; don't put them down
- be aware that vulnerable parents are likely to take your treatment of their daughter personally, as if it's directed at them; respecting every girl in your care contributes to respectful home-school relationships
- send home good news, not just bad
- at consultation evenings, take parents' concerns seriously, and end with: 'Is there anything else?'; an alternative time for a discussion can be arranged if the issue demands it

72 Channel competition creatively

Most girls have a competitive urge. Channelled sensibly, sensitively and creatively, it can be used to develop motivation, achievement and a positive sense of self. Exploited carelessly, it may lead to anxiety, despair, doubt and a decision to opt out of trying to do well.

Girls do best when they are encouraged to compete against themselves, when they focus on improving their last best performance, because this keeps their self-esteem intact. Competition is potentially dangerous when girls perform to impress their friends or adults, set their sights on beating others, and invest their self-worth in the result. Even though a girl may have tried really hard and prepared well, others may have tried harder, or simply have a more natural talent. Whenever the result is not wholly within her personal control, there is the chance that anxiety will threaten a girl's self-esteem and increase the likelihood of a poor performance.

Parents

- it is better to identify a specific target ('Try to improve the mark for your piano scales') than a general one ('Go for a distinction this time')
- don't fuel competition between brothers and sisters; each child needs to be successful in her own way, and to be accepted unconditionally for who she is
- avoid competing against your daughter, especially if you intend it as a spur to greater effort
- fun competitions are fine: 'See if you can beat me to the top of the stairs' is a great way to get children on the road to bed

Teachers

- research shows that classrooms run on competitive lines produce anxious children
- encourage pupils to perform in order to improve, not to impress, and give them feedback, too, so that they can see the progress they have made
- co-operative games enable children to have fun without any need for winners or losers; try playing musical chairs in two ways: first, as usual, removing chairs and children each round; then removing only chairs, which means that children have to sit on top of each other; ask which version they prefer

73 Failure lights the route to success

All girls experience failure – lots of it. A girl will almost certainly fail at her first attempt at walking; she won't master buttons on her first fumble, ride a bike or tie a bow straight away, yet she is prepared to have another go. Why don't these early failures make growing girls give up, despite sometimes intense frustration, while later ones can stop them in their tracks and throw them into the depths of misery?

The uncomfortable truth is that adults are often responsible for the change. They start telling children off for failing, tease them and make them feel ashamed.

Yet failure is not something to be shunned. It provides neutral information about what went wrong and what needs to be put right. Failure is an inevitable and essential part of learning, and shows that learning is happening at the frontier of current knowledge. If the lessons that underlie failure are taken on board, the errors light the path to success. But it won't happen if adults deny, ignore or punish failures, making a girl feel she needs to cheat, hide the truth or run from it herself.

Parents

- respond constructively – failure is like a puzzle to be solved, not a disaster to be denied; consider whether the target was too ambitious
- respond genuinely – be honest, and ensure that it remains your daughter's problem, not yours
- respond sensitively – however much she may deny it, failure is upsetting and can undermine confidence; accept, understand and let her voice her feelings; don't be too strict with her for a while, and help her to feel successful in other ways
- show that you love her for who she is, not for what she can do
- don't punish her for failing – she may start to lie or cheat

Teachers

- describe in detail what went wrong and how she may do better
- let the student know that you believe she can improve; show her how with a sample piece of work
- encourage her to self-evaluate as much and as accurately as possible
- be available if she needs help
- look for modest, not exceptional, success stories where individual girls can explain how, and perhaps why, they overcame failure
- try to discover the reasons behind any unexpected fall in performance

74 Watch her doing something she enjoys

When I got older, my mum didn't come ice skating with me anymore because I went with my friends. I carried on getting better, then one day I made her come again because I badly wanted her to see how well I'd got on. I'd started from nothing and I was so proud.

Anything that is important to us, we want to share. Children are the same. Children usually enjoy doing the things they do well, so one reason to watch your daughter is to let her show off a little and accept the pride she feels in her achievement. Watching brings you further into her life; the simple fact that you are there increases togetherness and helps her to feel comfortable with herself.

Whether your daughter enjoys rollerblading, ice skating, conquering playground equipment, dressing up, acting, computing or playing a musical instrument, watch her; it may take a bit of effort, but she will thrive in the warmth your sharing will bring.

Parents

- find the time to watch your daughter
- 'Mother, can I show you how I...?' or 'Dad, come and see me do...' should be answered: 'Sure' (preferably not with 'later' added)
- let your daughter impress you; tell her afterwards how much you enjoyed watching her
- tell somebody else, such as a grandparent, how well she's doing, within her hearing, so that she can feel proud of her skills; if no one's available, you can say: 'Your Gran would have loved to see you do that'

Teachers

- you can't watch what the girls in your class do in their personal time, but you can find a moment to hear about something they have done
- try to value whatever it is students enjoy doing or achieving
- provided that family and work commitments allow, it can be gratifying for students to see other teachers attending musical or sporting events

75 Don't invest your self-worth in her success

Every time I did well at something, my mum would rush off and tell the whole neighbourhood. She treated it as her achievement, and I ended up feeling used, empty and angry.

When a parent's self-esteem depends on, or is disproportionately enhanced by, their daughter's success, this imposes upon her a heavy responsibility that no child should have to bear. Although parents and teachers naturally feel good when their daughters and students do well, it is very dangerous when adults begin to rely on a child's achievements for their own sense of self-worth.

It can damage a girl's self-esteem in a number of subtle ways. If parents feel good about themselves only when their child succeeds, they are, in effect, stealing success from that child. This will leave her feeling used, confused and empty instead of fulfilled. Only further, and repeated, success will replenish her sense of achievement, which leads to the burden of perfectionism. She will also come to believe that she is valued solely for what she can do, not for who she is.

Parents

- if you want to tell other people about your daughter's achievements, think about who you want to tell, consider why, and ask for her permission to do so first
- avoid setting your daughter a new target as soon as she has reached one (Might you benefit in some way from this pressure you put on her?)
- consider whether you have different expectations for your sons and your daughters, and whether you identify more closely with one or the other for some reason
- tell yourself firmly that it's her success, the result of her effort, and hers to hold on to

Teachers

- good teachers deliver more than results; to stop yourself focusing too narrowly on them, at a time when there's every incentive to do so, list other ways in which you are keen for your students to develop and achieve
- if you think you may have become success-dependent, reflect on the other things you are good at, and which give you pleasure
- if one group's results are not good, and you find yourself getting depressed as a result, put yourself back in control – consider what you could do differently next time that might change the outcome

76 Let her be responsible for her own success – and failure

The moment of victory is far too short to live for that alone.
– Martina Navratilova, former tennis champion

Success and failure tend to be overlaid with moral significance – it is good to succeed and shameful to fail. Parents can be affected too, so that if their daughter succeeds, it is their success, and if she fails, it is their failure. This tendency to take ownership is damaging as well as confusing.

When parents take responsibility for a girl's success, it is effectively stolen from her and it may lead her to fail in the future. A mother might appropriate her daughter's success to make herself feel good, and run to tell others; or she might take credit for it, implying that it would not have happened without her input. If her success is always taken from her, a girl may eventually turn on her tormentors and refuse to play, or she may burn out.

Taking responsibility for a daughter's failure is equally unhelpful. Your shame may lead to punishment or trivialisation. Either will prevent her learning from her mistakes and she won't progress.

Parents

- have realistic expectations of your daughter and accept her unconditionally
- see success as neutral feedback; you can acknowledge it, but glory shouldn't come into it
- help your daughter to feel comfortable with her feelings of delight or disappointment, frustration or sadness
- if her 'failures' become your personal shame, you hinder her chances of recovery
- mistakes are a sign that she's at the frontiers of her knowledge; discuss why they happened and what she can do differently next time
- your daughter's success is hers, not yours; don't claim the credit

Teachers

- always congratulate a student on her success, and hand her the credit
- explain in detail what she has done right, so that she knows how to repeat it next time
- offer her some unpressured time at her new level, time to adjust to and accept her success; then wait to see whether she progresses on her own initiative
- take the shame out of failure; personal ridicule will encourage students to hide behind excuses

CHAPTER 8

Encouraging Confidence and Independence

We are moving from a 'to be or not to be' generation, to a 'to have or not to have' generation.

– Trond Waage, Norway's Ombudsman for Children)

True happiness, it has been said, is not a place of arrival, but a manner of travelling. Whether you have it depends upon how comfortable you are with yourself; this is the inner confidence that is called self-esteem. 'Happiness is self-contentedness' wrote the Greek philosopher Aristotle three centuries before Christianity.

The consumer society may give children injections of pleasure, but it tends to reduce the opportunities they have to discover who they really are. The more a young girl follows fashion or a group, the less certain she is about her own preferences or inclinations. The less she looks over her shoulder to check her acceptability to others, the more confident and independent she will be.

Confidence, of course, ebbs and flows. We can feel confident in some situations and terrified in others, relaxed with some people and uncertain with others. Confidence also dips naturally at certain times as part of normal child development. Girls' self-esteem plummets at 14, the age at which boys express themselves most freely, and boys reach their lowest level at 19. *The Can-Do Girls* and *Leading Lads* (see page 224) are two studies of girls' and boys' confidence and self-esteem. They found that while, overall, fewer girls express confidence than boys (21 per cent to boys' 25), more girls than boys were in the middle or very confident bands, and fewer girls than boys had low self-confidence (8 per cent to boys' 12 per cent).

A girl with confidence will know her own mind, honour and care for her body, be aware of her capabilities, have trust and faith and the ability to give pleasure to others. If she can make sense of her immediate world, she will be able to go out into the wider one with hope, purpose, passion and direction.

Confidence does not come from believing you are perfect, but from knowing you are good enough but have more to give. If you like enough of yourself, you are able to live with the bits that are less endearing. When you know you are good at some things, you can acknowledge, with no loss to your self-image, that you are not so good at others, and you can face the world with honesty and humility.

Your daughter will grow in confidence and independence if you first show confidence and trust in her, offer her security and certainty, and give her time and attention so that she can believe in herself.

77 Offer her safety, security and predictability

Confidence is defined in the *Concise Oxford Dictionary* as 'firm trust; assured expectation; self-reliance; boldness'.

Nobody can develop self-confidence if they neither trust themselves nor have the assured expectation that other people's behaviour is trustworthy and predictable. Girls who do not have a measure of consistency and predictability in their lives will find it hard to acquire the necessary trust, in others or in themselves, to become either truly self-confident or genuinely independent. When adults behave in an arbitrary and neglectful way, they undermine a child's confidence and generate emotional dependency.

Routines help to create both trust and security. If a girl's key carers clearly trust her and provide consistency, she can begin to trust herself, her judgement and the behaviour of other people. Parents don't make their daughter independent by disappearing from her life and letting her fend for herself, but by being there for her.

Parents

- try to ensure that your daughter begins each day with a clear idea of what will happen and when, either in terms of routines and events, or who she will meet
- as far as possible, arrange for her to see any non-resident parent regularly and reliably
- make your own behaviour towards her as reliable and predictable as you can; if you have moods, or your routine has to change, try to explain why
- consider whether you could change your own or her commitments to increase the sense of pattern and 'assured expectation' in her life

Teachers

- all children feel safer when lessons have a clear structure and purpose, and aims and objectives are made clear at the start
- clearly structured tasks in lessons will encourage students to speak and listen in a purposeful way, and will help them to gain confidence in thinking constructively and creatively
- let them know well in advance if there are to be any changes to the normal school routine or lesson

78 Nurture her social skills

Girls, typically, not only talk more than boys as they play, but seem more ready to make and trust friends. Their brains also give them a natural verbal advantage. All these things encourage good social and communication skills.

These skills help to give girls a secure and happy future, socially and economically. Employers are increasingly looking for people who are polite and respectful, can work together in teams, reflect on, discuss and sell their ideas, and manage disagreements with skill. Young people who grow up able to share, compromise and be flexible, and who have an understanding of how others think and feel, will have a head start. They are also more likely to have happier and longer-term personal relationships, which so often founder when differences are not managed either constructively or safely.

The best way to help your daughter become socially confident is to ensure there is plenty of fun, conversation, discussion and respectful listening within the family.

Parents

- involve your daughter in your social life whenever appropriate; it will show that you enjoy her company and trust her to behave well
- try to be sociable yourself; if she sees you mixing confidently with others, she'll learn from you
- encourage friendships; invite friends round, and see if there's a social or sports group she can join; solitary activities, such as computer games, should be rationed
- reading will develop her reflective powers and insight; talk to her as much as you can: ask her for her views, tell her what you've been doing – and listen to her properly when she talks to you.

Teachers

- include paired and small-group work in your lessons so that pupils learn to listen, talk to, respect and compromise with others
- in mixed classes, ensure that girls work with boys, so that they pick up alternative ways of thinking, doing and learning
- seating arrangements are important – it can matter a lot to pupils who they sit next to
- encourage discussion, listening and reflection; when children get into the habit of thinking before they write, the standard of their work rises significantly

79 Offer chances to test herself

Confidence flows from competence. When a girl possesses a variety of skills that she can rely on in different situations, she will feel confident and capable. She will expect success instead of anticipating failure when faced with obstacles.

The more skills she acquires, the more competent she will feel; but she won't become competent at anything if she watches a great deal of television or is told endlessly that she's no good when she does try something. Beware gender stereotyping when you recommend activities for your daughter to try: she should spend plenty of time outdoors, playing and running about, to develop her strength, co-ordination and long-term health.

At the right time, a taste of the world of work may help her to become confident about her aims. Schools increasingly organise this, but further experience can only be beneficial, especially if she can join you and appreciate the responsibilities you have outside the home.

Parents

- every journey begins with a first step; your daughter won't be as good as you, but she'll need to feel competent from the beginning; teasing her for failing will undermine her confidence

- activity holidays and after-school clubs can introduce girls to a range of new skills

- involve your daughter in the jobs you do – cooking, cleaning, car washing, weeding, DIY or doing your business or home accounts – and be patient if they take longer to do

- let her try out her thoughts and judgements on you; invite them, hear them, and respect them; don't dominate or compete

Teachers

- girls relish responsibility and like to seize the opportunity to take it; make sure that the boys in your class are not left out

- encourage a girl who seems to lack friends, confidence and social skills to attend any after-school clubs available

80 Work to and from her strengths

Any girl will learn more easily, perform better and be more self-motivated if she can do things in a way that suits her and interests her. Some children learn better by looking at visual images, some when sounds are used, and some through touch. A child's all-important sense of mastery – how effective and competent she feels – will develop best when she is allowed to start from who, and where, she is.

Success breeds success. All educators know that when a child succeeds in one field, she will have a more positive attitude towards the next challenge she has to face.

When you can acknowledge your daughter's strengths, she will feel not only understood but also accepted, which will free her to learn and develop in her own way. The more you impose your own methods and ignore hers, the more likely she is to lose the confidence and ability to work independently.

Parents

- identify your daughter's skills and strengths: concentrate on what she can do, not on what she can't
- don't belittle the talent that she feels is special to her
- think about her preferred approach to working and learning, and her particular passions; don't force her to work in ways that she finds difficult

Teachers

- discover her passion, apply it to your subject, and let her fly; all children concentrate better when they're confident and committed
- quizzes make learning fun, and help girls to become more comfortable with risk-taking and uncertainty
- young girls' play is often less vigorous, and more verbal and fluent, than boys'; but there are always quiet boys and boisterous girls who cross the boundaries; don't label girls who prefer rough and tumble as 'tomboys'

81 Give her independence, but don't abandon her

Independence and responsibility should be given to children as and when they are ready, and not be granted solely because it is convenient for an adult to do so. The child's age, maturity and wishes should be taken into consideration. Sometimes, a family's practical needs are the trigger for granting further independence, and it comes at the right time. However, it is important to judge whether giving your daughter extra freedom or responsibility is simply a convenience for you, and to be aware of your daughter's possible view of your motives. She should not be given too much too soon, be exploited, or be left feeling abandoned. She may appear to be able to cope, but in reality she may still feel she needs your company, guidance and attention, but be too proud to say so. And if she becomes anxious or feels out of her depth, her confidence will be weakened, not strengthened.

Parents

- for the first few times your daughter does something new, stay close by or within easy reach, then she will know that she's not entirely on her own
- apply the idea of 'supported independence' to assess whether she might feel abandoned or neglected; check whether her friends are available and reliable for times when you cannot be contacted to provide support if needed
- if she has done something on her own, ask for her honest opinion about the arrangements you made for the resolution of potential problems or in preparation for her travelling, being or coping on her own

Teachers

- 'independent learning' is a valuable approach, but students will continue to need support
- girls may need help with time management; for projects with a long deadline, arrange to be available for consultation at scheduled times well before the due date
- offer 'bite-size' chunks: suggest interim deadlines for sections of work to prevent any student falling behind
- tasks and responsibilities should not be granted without clear guidelines
- when students work on their own, encourage them to think ahead about areas where they might need help

82 Monitor and supervise, from a distance

One night I was due home by 11.00. My friends and I cooked up other plans, so I phoned home with some made-up story about why I needed to stay the night with one of them. Mum just did not buy it and stuck to her guns. Afterwards, I was glad. I realised I wasn't sure what might have happened that night.

Research shows that many girls who get into trouble with the police are out for long periods of time without having to report their comings and goings to anyone at home. Monitoring and supervision help to keep girls on track when they are exploring their new-found freedom, but parents should watch over their daughters sensitively and from a distance, so that they don't feel insulted by any perceived lack of trust.

Keeping yourself informed of your daughter's movements is vital. It helps to keep her safe, because she knows that you're aware and that you care. If you cast her adrift, she may flounder, feel neglected and then get her own back by seeking out trouble and mischief.

Parents

- discuss and agree with your daughter times for phoning and coming home for every outing; afterwards, casually ask what happened and how things went
- if you are worried, check with one of her friends' parents that they were told the same story – make sure you have these contact numbers
- if she's late, find out why, so that she knows you notice, and care; if possible, wait up for her and see how she is
- take a look at the club or place where she goes, to get a sense of its atmosphere
- keep an eye on her bedroom in case it gives clues about any work or personal problems

Teachers

- be alert to any pattern in missed deadlines; consult with colleagues if you are worried
- patrolling children closely won't help them learn to manage and monitor themselves; keep an eye open at break-times, but from an appropriate distance
- it is easy to keep records of students' marks for class and homework, of late arrivals, and of major behaviour incidents; but it's important to keep track of other less obvious things which may indicate personal trouble, such as tears, visits to the school nurse and changes in personal appearance

83 Develop responsibility and safe risk-taking

From birth onwards, research shows that girls are more placid than boys, more thoughtful and cautious and less inclined to take risks. They are also more likely to try to please, being in general compliant. These tendencies may make them easier to manage when they are younger. However, as they approach adolescence, their stronger attachment to their family can make it more difficult for them to achieve independence. This may explain why girls seem more prone to the 'tempestuous teens' than boys.

When girls are given freedom all at once, or where they have torn themselves free in anger from families that have restricted them, their risk-taking can be dangerous. Of course, our daughters may get into trouble even if we have, we believe, prepared them well: for the more freedom they have, the bigger their potential mistakes. Nevertheless, the best preparation for safe risk-taking we can provide is to let them know we care, to treat them fairly, to encourage their thinking skills and self-respect and, crucially, to see that their responsibilities grow in proportion to their rights.

Parents

- when your daughter asks for more independence, try to give it to her; if you feel the particular freedom she seeks is not appropriate, discuss an alternative change that will satisfy her – she'll then have less need to struggle free and prove herself in irresponsible ways
- be tolerant of her mistakes; harsh punishments may make her behave more irresponsibly
- rights matched by responsibilities can encourage safer behaviour, but all girls will take risks at some point; discuss what you mean by safe risk-taking, and ensure that her freedom is set within clear limits

Teachers

- learning involves taking responsibility and taking risks; when kept in balance, this offers useful lessons for life
- challenging girls, or those with low self-esteem, may respond well to being given special tasks and more responsibilities
- girls who undertake death-defying acts of bravado can be mirroring the lack of care they perceive in close adults; be attuned to the reasons behind high-risk behaviour
- address safe limits to risk-taking in PSE lessons and assemblies; discuss the attraction of thrill and excitement and what these achieve

84 Encourage self-management

When the parents of our new Reception children first come into school, we explain why we like them to encourage self-care skills in their children. It's partly because we have so many coats to do up, but mainly because it has vital educational value. Young children who can manage themselves are more confident, responsible, independent and effective in their work.

It can be hard to let go. With such busy lives, we may feel we don't have enough opportunities to demonstrate our love and commitment. One remaining way, as daughters grow older and get harder to hug, is to tend to their needs. I met one very glamorous mother in her fifties who was still buying clothes for her two daughters, even though they were both away at university, because she felt uncomfortable about their lack of style. She was even happy to return the items if they were rejected. But it is not helpful to mollycoddle. It fosters dependency and will prevent a girl from learning to organise her work, time, money, and image, for herself.

Parents

- encourage financial independence; give your daughter regular pocket money or a clothes allowance, and stick to it
- encourage organisational skills, e.g.: if you're planning an outing, as a fun project, ask her to find out the opening times and costs; give her a budget and sole charge of the family purse
- even young girls can be encouraged to put on their shoes, wash their faces, clean their teeth and get their things ready for school by themselves
- don't back off altogether; do enough to let her know that you still care and think about her

Teachers

- make sure parents are aware that children who can look after themselves are also more successful learners
- don't be tempted to let a few competent students do all the responsible jobs; spread the tasks between all of them
- girls who live between two sets of parents may have difficulty remembering books, especially early on in the separation; it's better to give a child two sets than to scold her and add to her problems
- be rigorous in promoting the use of planners and homework diaries
- actively address time and stress management, work and planning skills

85 Let her say no

Self-esteem gives girls the power to say no – to their friends, or to an adult who is behaving in a frightening or strange way. Girls who are expected to be 'good' all the time and who rely on other people's approval to feel accepted, will find it much harder to withdraw from potential danger where this could result in being teased, told off or cold-shouldered.

Sound self-confidence is one route to staying safe, and parents and other carers will therefore need to nurture self-esteem as the core of confidence. But girls will also need a bit of practice. They can't turn from habitual 'yes' girls into fierce 'no' girls in one leap. Many parents may feel, not without reason, that their daughters are already too cheeky and willful, and need no further encouragement. But insolence and opposition are not quite what is required. What children need is being allowed to disagree; learning to stand up for themselves with reasoned argument, not fists; and knowing that their judgement is worthy of respect.

Parents

- let your daughter have views that differ from your own, so that she feels confident about being in a minority of one when necessary
- allow her to express her feelings; if it's OK to feel angry, sad or excited at home, it will be easier for her to respond honestly and decisively in potentially dangerous situations
- let her know you trust her judgement
- listen to her properly when she wants to tell you something
- respect her choice of friends wherever possible; if you criticise too often, she's more likely to ignore you when you have serious concerns

Teachers

- employers no longer need armies of automatons or lines of lemmings; ritual obedience is out- of-date, though politeness and respect certainly are not; older children, especially, must be given the space to make up their own minds
- listen; you don't have to give in to show respect to a pupil's right to see things differently and say so, provided that she expresses herself politely
- explore the scope for compromise; she may have a good case, which can be answered in another way

86 Teach coping and survival skills

The best way to learn to manage many problems is through experience. Experience lessens fear and also helps to build common sense. Hiding from fears makes them grow. Staying at home or in a car does not build life or street skills, and wrapping girls in cotton wool is not responsible parenting. Getting out and about with your daughter, walking, cycling and going on public transport teach road and geographical sense and street awareness; going out for night walks will help her to respect, but not fear, the dark.

Staying indoors does not encourage physical fitness. The two best defences against bullying and other dangers are strong bodies and inner confidence. When girls are fit, physically strong and have good posture, it helps them not only to run away or wriggle free but also to look confident and convey to others they are someone who is best left alone.

Discuss different coping strategies with your daughter, particularly ways to reduce risk and protect herself in potentially dangerous situations.

Parents

- useful risk-reduction advice includes: stay in public view and in populated places; avoid back stairs and subways; go out in a group – preferably composed of people you know and trust, and don't become separated from them; carry money safely – with a small amount in a purse and the rest elsewhere
- practise your verbal responses; a sharp word delivered quickly is probably safer than a punch
- before she goes out, ask your daughter if she's worried about anything; if she is, discuss it
- it's important that she feels confident; don't undermine her by raising the spectre of disaster
- if she's going out drinking, give her a good meal first

Teachers

- if it is practical, organise a termly 'walk-to-school' day or week
- include safety and survival issues in PSE lessons, but keep all discussions as positive as possible: fear of 'stranger danger' can get out of hand and destroy children's confidence

87 Enter the no-go areas

Sex, alcohol and – increasingly – drugs in every form are an inevitable part of growing up for girls today. These topics should not be allowed to become no-go areas in your family. Though your daughter needs her private space as she enters her teens and will defend this sometimes aggressively against your intrusion; and even though it's hard to strike the right note and avoid embarrassment, it is important to keep talking. Communication must be maintained, so that if serious problems arise regarding sexual behaviour or alcohol of drug abuse, you can address these straightforwardly, because the territory is familiar.

By the age of 14, research tells us, one in three young people will have tried at least one form of recreational drug. If you don't talk to your daughter about drugs, someone else will. Telling a girl not to do something when peer pressure is strong may not have much impact, but suggesting that she stays in control of, and true to, herself by doing things only when it feels right for her may give her that ounce of extra courage to say no.

Parents

- at home, try to talk openly and comfortably about sex in general conversation, so that the subject is not unfamiliar or taboo
- if it has become difficult to talk to your daughter, leaflets and books about safe sex, safe drinking and drug use are available from the Family Planning Association, health centres and schools
- be quietly vigilant; become informed about and watch out for signs of inappropriate behaviour or use (too much or at the wrong time) of either drugs or alcohol
- if you begin a new relationship during your daughter's puberty, be aware that she may find the sexual side of it problematic; be discreet, don't compete

Teachers

- most schools, both secondary and primary, have sex and drugs education programmes within an increasingly coherent PSHE curriculum; these need to be given high status, and presented by a qualified staff member: someone with the right interpersonal and professional skills and knowledge
- teaching should reinforce high self-esteem and good social and communication skills as the best defences against premature sexual activity, early pregnancy and drug use
- PSHE teachers should possess excellent group-work skills to enable all students to participate comfortably and confidently and take the subject seriously

88 Promote self-direction

People don't resist change, they resist being changed
– Gerard Nierenberg

One form of independence is self-direction. Children who are self-directed know what they want to achieve and can knuckle down and manage tasks and problems independently without needing to rely on adult supervision. This helps them to experience autonomy, because they have sufficient control over aspects of their lives to feel powerful (not passive) and what is sometimes called 'authenticity' so that they are able to make their actions follow their thoughts. Girls who are given no chance to direct themselves, or who lack the skills or confidence to do so, feel helpless. And when they feel helpless, they soon feel hopeless.

Self-direction and independence reinforce each other. The more self-directed girls are, the better they can manage independence; and the more independence they are given, the more likely they are to become confident and self-reliant, to show initiative and be creative. However, make it clear to your daughter that although you are promoting self-reliance, this does not mean she can't come to you if things go wrong and she needs advice.

Parents

- directive parents create dependency; the more you tell your daughter what to do, the less competent she will feel and the more she will need you to do it
- give her pocket money as soon as she can manage it, then she can spend without reference to you
- when children feel helpless, they soon feel hopeless
- if you feel that your daughter should change in some way, involve her in deciding when and how; iIf she wants to change something, co-operate
- to become self-directed, girls need discretionary time; filling your daughter's every moment will deny her this

Teachers

- students improve when they understand how to make progress: set clear objectives and targets
- a pupil will feel more in control of what she has to do if you ask: 'Do you want to do it this way, or that way?'
- once she has a target, ask her how she plans to reach it
- encourage her to think and plan ahead and manage her own time, meeting her needs and yours
- encourage self-appraisal as part of the process

CHAPTER 9

Checking Out Your Role and Feelings

This is where we come back to basics. Despite a growing emphasis on both the power of genes and the acknowledged pull of the peer group, parents and other key caring adults are in a strong position to influence a girl and to affect how much faith she has in herself, how competent she feels and thus the overall quality of her self-esteem. A child can be born with a predisposition to be positive about life and herself, or with a negative tendency. Close adults can either build her up or undermine her. A family may have two daughters who couldn't be more different from each other, but one may need a great deal more support than the other. It is our role as adults to provide, as far as possible, the conditions within which each one can feel secure and capable, not uncertain and incapable. She must be able to influence her life, not merely react to events and play the victim. She needs values, a direction and the capacity to enjoy activities, causes and people, not to be rootless, isolated or completely self-absorbed.

We can influence, but we cannot control all outcomes or even always manage our own behaviour as we would wish. Outside factors intervene. Stress, uncertainty and change prevent us from giving as much as the job of parenting sometimes. Frequently, our daughter's behaviour will test us to our limits. She contributes to the relationship dynamic, too, and, as she approaches adulthood, she must be held increasingly accountable for her own behaviour. If however we have provided most of the basics, accepted her imperfections as well as our own, demonstrated our commitment in a way that satisfied her, and bolstered her confidence whenever she became vulnerable, we have given her a firm foundation which will help her to cope confidently with any future setback.

Our feelings, hopes and fears, inevitably colour what we say and do. No human being is sufficiently saintly to think about someone else's interests all the time, to the exclusion of their own. Children demand, and need, a great deal of time and attention. Giving as much as they sometimes want takes a lot out of parents. Indeed, they can take so much out that you may wonder whether you have anything left. If you do not consciously take the time regularly to replenish yourself, recharge your batteries and develop your own sense of self-esteem, you may find you build protective barriers around yourself in a desperate and arbitrary way. The barriers may help you to hold on to yourself, but you may also cut yourself off from your daughter when she needs you most.

The best way to help children grow up happy and healthy is to make sure that you also continue growing and enjoy your life within and without your family.

89 Cherish and trust yourself

At the same time as looking after your daughter, you have to look after yourself. You probably don't need to be told that the better you feel about yourself, the better you cope with challenge and difficulty and the nicer you are to those around you. You will know, too, that when you have had a bad day or are very tired, you are more likely to 'take it out' on your nearest and dearest. Looking after yourself is an investment which benefits others, for when you behave well and notice good things that others are doing, you help them to feel as good as you do. Good behaviour is infectious.

It can be hard to trust your competence during every stage that children pass through and to manage every issue that arises. Most parents and teachers have a 'favourite' age for children that they enjoy more than others. While teachers can choose whether to teach little ones or older children, parents have no choice: they have to cope throughout. You may often doubt yourself, but remember that children value firmness. Discuss any uncertainty with others, carefully review your first reactions, and if you still feel the same, trust your own judgement.

Parents

- talk to others: it usually helps; there may be a parenting group near you to join
- identify your little luxury, the thing that calms you down and restores your faith in yourself; it may be reading a trashy book, going to the pictures or having a drink with friends
- make sure your choice of 'pick-me-up' is realistic; when grandiose schemes fail, it can have the opposite effect
- try to treat yourself on a regular basis; while some little 'indulgencies' can be fitted in easily, others take longer and require planning
- 'I've always wanted to...', so do it

Teachers

- trust and believe in yourself; if you doubt your skills, you may interpret students' difficult behaviour as a personal attack, and react defensively, provocatively and unconstructively
- list what you see as your professional strengths, then identify where there's room for improvement; discuss with colleagues how to share your collective skills, to aid professional development
- managing girls who challenge is tiring; rather than suffering in silence and pretending you're coping, set up a support group with colleagues to pool ideas
- after a bad patch, pamper, don't punish, yourself

90 Let her be different

I can't believe how different I am from my mum. She was a piano teacher, a bit brainy, and loved endless walks. I hated classical music, didn't like reading or school, and those walks were just awful. My brother and I were dragged along, but as soon as we could, we were allowed to stay behind. She also let me give up the piano after two terms. It was such a relief.

We spend the early years of our child's life treating her as a mirror – looking for ourselves in her. It starts with her face: 'She's got my eyes/her dad's nose.' Then you move on to her likes and dislikes: 'She loves tidying her cupboards, just like I did.' All the similarities are proof that our daughter is part of and belongs to us; so it can come as a shock later on when we are forced to accept that she is not only different, but also will go out of her way to prove it.

If you use your child to make yourself feel acceptable, she will feel shackled and stifled. Growing up is hard enough, but it will be far harder if you make your daughter responsible for your own happiness.

Parents

- give your daughter the space to be herself; it's your problem, not hers, if you feel uncomfortable about her being different

- don't compete with, criticise or ridicule her; every challenge drags her onto your territory and reinforces her view that you think your ways, talents and preferences are better than hers

- getting involved in her school is great, but don't overdo it; all girls need some space to be themselves, free from a parent's watchful and expectant eye

- take an interest in the music, magazines, games or clothes your daughter likes, even if each is only a passing fad; you're not expected to agree – it's her choice

Teachers

- encourage children to be aware of their individual likes and dislikes, temperament and idiosyncrasies

- explore class activities that highlight and develop respect and tolerance for differences

- vary your teaching methods; the way you feel most comfortable putting information across may not suit the learning styles of all the girls in your class, and they may be too polite or intimidated to tell you

91 Inspect your expectations

Every family has its own story and preferred way of organising itself. These influence what we expect for ourselves and those we live with, including our children, whether we want them to or not. There is often a hidden, sometimes complicated, agenda, which children eventually detect and react to.

While we accept that education and careers are important for girls, there are many families in which women still play a more traditional role. Some mothers will want their daughters to grasp the opportunities they never had, and some will prefer them to follow in their own footsteps. Where parents' expectations differ, a girl may become further confused about the course she should take in life, especially if she usually pleases them, not herself. There is some evidence that confusion about identity may contribute to problems such as eating disorders.

Some typical stories behind expectations include: you threw away the opportunities you had, and you don't want your daughter to do the same; you are successful, and see her success as another feather in your cap; you want your daughter to do what you always wanted to do, but couldn't.

Parents

- list the expectations you have for your daughter; try to be honest about why you hold these, and think of the possible positive and negative consequences for her of each one
- think separately about your daughter's sporting activities, music, art, school, career, hobbies and leisure, and whether your expectations are high, medium of low for each one; if there are lots of 'highs', think where you might lower your sights
- ask your daughter whether she agrees with your targets and whether she feels she can, or knows how to, meet them
- looking at just your short-term academic expectations, consider whether they are realistic, and how you will feel if she doesn't meet them

Teachers

- girls benefit from being set high, but achievable, short-term goals; be careful that your personal goals and school targets don't put undue pressure on your students
- help each pupil to define her own expectations, and step in only if she has made a serious misjudgement

92 What you expect is what you get

Time for school in 10 minutes. Remember what you need and I'll see you at the front door at quarter to.' For older girls, this is a far more helpful and positive approach than: 'You've only got 10 minutes. Have you got your homework? Have you done your teeth? Don't forget your games kit, and don't be late like you were yesterday.

Girls who feel trusted by an adult feel proud of that trust, and work at keeping it. They live up to expectations to strengthen it further. Research shows consistently that high expectations of behaviour produce good results, and low expectations produce poor ones. 'What you see is what you get' is shorthand for this process. 'What you see' in this case is your perception of your daughter's personality and behaviour. 'What you get' is the behaviour you expect. So if you ask a girl to do something in a way that assumes and expects that she will comply, you are more likely to get the result you want. The reverse is also true; when you let slip that you think your daughter will not co-operate or succeed at something, she probably won't.

Parents

- stay positive: notice, constantly, the things that your daughter does right; if she fails to do something, restate your request or expectation, don't berate her
- stop predicting or assuming poor performance or behaviour with phrases such as: 'I expect you'll fail this test, too' or even 'You will be good, won't you?'
- avoid sticking labels on her, especially negative ones such as 'naughty', 'cheat', 'liar', 'hopeless'; give her hope and faith in herself
- beware of asking too much of her, and thereby becoming her central reference point

Teachers

- have appropriate and realistic expectations for content and presentational quality of students' work, and high expectations for prompt delivery
- many girls take a pride in producing neat work, but obsessively 'manicured' assignments can imply a dangerous perfectionism; encourage risk-taking, and trial and error, rather than perfectionism
- many girls are comfortable with private reflection, but fear public discussion; class discussion will help them to become more confident about shaping and expressing their ideas

93 Watch what you say

Words are enormously powerful. What parents and teachers say, and how they say it, can have a far greater impact than most adults realise. I know of one girl who became dangerously thin because her teacher described her as 'the fat one', comparing her to her thin sister.

Without realising it, we can say things that put girls down, humiliate them and damage their self-belief and self-respect. Fathers and father figures are especially prone to engaging in playful, teasing banter with their daughters, using threats, sarcasm and insults as a way of showing love. They may do this partly because straightforward praise and intimacy are strange and uncomfortable to them. However, vulnerable girls will never be certain that no real criticism was meant.

If we want to help our daughters feel loved and cared for, we must try not to make comments or give reprimands that undermine them, even in jest.

Parents

- verbal teasing is a form of manipulation that should be used carefully and sparingly
- be positive; give plenty of praise and kind words; these won't make your daughter conceited if you teach that 'good at' means 'different from', not 'better than' in personal terms
- be aware that when we put our children down, it can be a defence mechanism to protect our own sense of inadequacy
- phrases such as: 'I can't take you anywhere', 'You'll probably end up pregnant', 'I don't care what you think', 'What's so good about that?' and 'You'll never learn' will progressively destroy your daughter's self-esteem and self-belief

Teachers

- be positive at all times; careless and insensitive remarks are taken to heart more than you might imagine or intend
- research has shown that confident five-year-olds entering school can become uncertain and develop 'learned helplessness' when subjected to constant criticism and negative comments about their work and play
- it is good for girls to be active as well as to sit quietly; be sure that your comments don't reinforce acquiescence and discourage more adventurous pupils from exploring

94 Loosen the straitjacket

Why are you always so lazy and messy? Why can't you be tidy like your sister? She just gets on and puts her things away with no arguments. You're just like your father!

'Straitjackets' are statements that lock a person into a role and deny any possibility of change. 'You always...,' 'You are just like...,' 'You will never...' are typical straitjackets. We all keep developing until the end of our lives. It is unfair in the extreme to hold fixed ideas about anyone, especially a child. Straitjackets can encourage your daughter to become whatever it is you say she is, because she'll give up trying to persuade you that you're mistaken.

Straitjackets come in two forms: labels, insulting and hateful jibes that describe what a girl is and what she is not ('You are a moron, an idiot', 'You'll never be any good at school', 'Why don't you ever tell the truth/finish anything?') and comparisons which unfavourably measure a girl against someone else ('Ahmed's much more reliable than you').

Parents

- 'Why are you...?' accusations are the most insulting, because they force your daughter to acknowledge your description of her in order to defend herself; 'I find you ... when ...' is more acceptable, as it puts the emphasis firmly on your feelings, is specific about the time, and makes clear that she's not always like that
- try to move from 'Why?' questions to 'I' statements
- make a mental or written note of the things your daughter 'always' does, then be on the look-out to notice when she doesn't conform to the pattern
- take the accusation 'always' out of your vocabulary

Teachers

- don't compare; when you find yourself teaching children from the same family, never mention the talents or failings of one child to the other
- try a class activity in which the students discuss different types of insults and the impact they have; blame, sarcasm and ridicule are forms of put-down or insult that can be considered; your students can practise using 'I' statements instead
- unbuckle your own straitjacket; teachers can be as guilty as parents of failing to revise opinions about individual personalities in the face of evidence to the contrary

95 Model respect for women

Mum, why did you let Dad say those things to you and not answer him back? It really upset me to hear it. You must be weak inside'

The best thing a father or father figure can do for the children in his care is to show love and respect for their mother. This reinforces her authority, strengthens the children's respect for her, and makes them feel integrated and secure in themselves and in the current family set-up. Crucially, it also models respect for all women. When girls grow up seeing their mother respected, it helps them to respect themselves and feel good about who they will grow up to be.

If we want to encourage girls to develop a more self-regarding image of femininity, we must model respect for all women, including women teachers and drivers, and the particular strengths and features that women possess. Verbal or physical abuse and violence towards women, and especially to a girl's mother, can be the single most damaging factor to a girl's self-esteem, her mental health and therefore her chances in life.

Parents

- don't let your daughters see your sons walking all over you; children won't grow up learning to respect women if their mothers don't respect themselves; taking time for yourself and maintaining house rules to protect your interests are marks of self-respect
- earn your daughter's respect; if you don't treat others, including her (possibly absent) parent with due respect, she might question your right to be respected
- watch out for the tone and content of your throw-away remarks when watching films or television or in the street; many terms of abuse imply disrespect to women

Teachers

- discuss whether you might introduce a sexual harassment policy in your school; girls should, of course, be prepared to treat boys with the same respect they would wish to receive in return
- gender awareness and equality of respect should apply throughout the school and in every lesson

96 Don't swamp her with your success

My dad was a successful, self-made man. He did it all on his own and he never let us forget it. He did really well, flashed his money around, and I had no idea how I was going to match up to him, which is what he expected. I was terrified when I left school and the future stared me in the face.

It's surprising how many girls follow in their successful mother or father's footsteps and do as well, if not better, in the same career. But for every success story, there will be another one, in which the daughter opts out because she feels she can't compete. The danger comes when either parent invests his or her self-worth in a child's success, perceiving this as the only way to gain acceptability. When this happens, the parents are forcing their values and their world onto their daughter, and she may have different plans and interests.

Parents

- manage your success with sensitivity; let your daughter see that it is the result of your interest, commitment and hard work, not your brilliance – and keep it as your thing, not something for her to marvel at
- be modest; your success will be evident – being a good role model doesn't mean blowing your own trumpet
- don't seek to excel at every activity; doing something well enough, without comment, and showing that you sometimes fail, sets a useful example too
- having two parents who are successful in different fields may make it harder for a girl to find her own niche; this makes it even more important to value the things she likes and does well

Teachers

- when a pupil is struggling with something she finds difficult, it won't help her if you show how easy it really is by racing through the calculation, explanation or arguments
- underachieving students may not feel capable of following in your footsteps, or in those of any other role model; explain carefully, therefore, each stage on the road to success, to make it appear manageable

97 What we fear, we bring about

My mum was terrified I'd get into drugs. She never trusted me, checked through my things, nosed into where I went and who with and she peered at me constantly. I got so fed up that I went out more, and got in with a crowd that was, yes, into drugs.

There's a frightening, almost magnetic, force that seems to operate alongside our fears: the more we want something for our daughter, or fear she will or won't do something, the more we seem to sabotage our best intentions. She senses how worried we are, and reacts in a negative way. If we're afraid she's going to grow up naughty, we use harsher punishments, which may encourage her to rebel. If we ban sweets and biscuits from the house, she'll buy them herself and binge when she's got the independence and money. If we force her to practise her music because she has the talent to excel, she'll lose her love of it and give up, and so on.

The common threads in all this are trust and power. If we have little trust, and we use our power inappropriately to manage our fears, we are more likely to turn them into a reality.

Parents

- identify your fears for your daughter, if any, and ask yourself whether you handle them in a way that may become counter-productive
- try to get your fears into perspective; discuss with someone else how real they are, how important it is that your daughter achieves in the way you wish and, if 'the worst' happens, whether it really will be that terrible
- cut out the power; give her as much scope as possible to manage herself, within your guidelines if appropriate
- reward the behaviour you want to see, rather than punishing the lapses, but don't be too manipulative

Teachers

- identify your fears for a particularly difficult class, individual or professional relationship; you may be afraid of appearing incompetent, disorganised or too harsh, for example; reflect on the pattern of your responses and how to change them
- think about the fears of the other person or people you have just identified, and how these might interplay with your own
- suggest that girls in your class be open about their own fears, in and outside school; they may fear that they are ugly, or appear 'uncool'; ask how their anxieties colour their behaviour

98 The more you use it, the more you lose it

I saw a young boy getting bored at the skirt of his mother as she talked with a friend in the high street. He decided to run away towards the road, so his mum grabbed him, dragged him back, hit him and resumed her conversation. He did the same again, twice, and so did she. The third time, he ran out into the road, and she hit him harder, several times. The more she hit, the more he chose to flout her authority.

There is an important lesson to learn about power, which is that the more you use it, the more you lose it. Where power is used or, more relevantly, misused frequently, it tends not to stop children doing something but to incite them to further defiance. Perhaps children see adults who have an over-reliance on power as weak underneath, and exploit this weakness. More likely, they resent the exploitation of their inferior status and skills, and not being understood or treated with the respect they deserve. They then express their frustration with the only power they have – to hit where it hurts.

Parents

- hitting is not the only power tactic used by parents; threats, bribes, harsh punishments and fancy arguments are also used by parents to get their way; children get the measure of these tactics too
- we remain authoritative not by being authoritarian and controlling towards our children, but by continuing to guide, influence, set boundaries for their decisions and, sometimes, to direct them if necessary

Teachers

- a girl who is confident will argue in order to make her point if she feels unfairly treated; if you fail to take note and listen to her, you could suffer the consequences later, when she becomes angry and frustrated
- if you relax your tight control, your authority will not automatically be undermined; research shows that if you put students more in charge of their own learning, and allow them to evaluate and modulate their own and each others' work under your guidance, the number of destructive, power-based challenges declines

99 Let her grow wings

Independence is a vital and exciting part of growing up for girls. Gaining in competence; experiencing challenges and surviving; experimenting with risks and different ways of doing things; gaining more control over what happens to her and learning to set her own boundaries – are all essential if a girl is to become an independent and responsible adult.

Unfortunately, the world outside the home seems to be an increasingly dangerous place. Parents are naturally worried about giving their daughters greater freedom to play and travel without adult supervision. Instead of encouraging their girls to go off on their own and experiment, giving out rope gradually, they tend to supervise, chaperone, constrain and contain them. Children are driven everywhere and discouraged from playing even in the front garden, let alone the street or park. When children are safe indoors, parents can relax. To add virtue to self-interest, they believe that they are doing the best for their child. But managing risk and coping with the unknown increase confidence and are important life and learning skills. Girls need to grow wings and learn to use them.

Parents

- opportunities to be on her own give your daughter a chance to test herself; without them, she'll find it harder to establish her identity, to develop self-esteem, and to achieve social adjustment (finding out how to behave in and belong to bigger groups)

- give her some freedom outside the home gradually, and at first insist that she remains with older siblings or within the safety of a trustworthy group

- don't go everywhere by car; let her travel on public transport, going with her to begin with, so that she learns her way around in the safety of your company

Teachers

- discourage parents from coming to peek at their child through the railings at break-time; if she spots them, their daughter could feel bound and restricted rather than loved

- some children learn to use their wings by trying out different ways of working; treat these trials respectfully, and be as flexible as possible

- class work in lessons such as Maths or Geography can involve the use of bus and train timetables, which will enhance self-management skills

100 Prove your commitment

Inner strength is built on commitment. Children need to feel they have the commitment of at least one significant adult in order to grow up happy, confident, secure and resilient, feeling that they have something to give to others. Birth parents are not the ones who can offer it. Girls who are able to commit to learning, organisations and friendships despite personal difficulties usually acknowledge the commitment of someone who spent time with them, showed interest in the things they did, accepted them unconditionally and, crucially, was reliable and there in times of need.

Your commitment will enhance your daughter's self-esteem, and help her to think for herself; to withstand the pressure of being in a minority of one, yet be flexible enough to compromise on issues of lesser importance; to listen to constructive criticism without perceiving it as a personal slight; and to have sufficient curiosity to learn, explore, think and look ahead. It will ensure that she has the power, and the will power, to determine and prepare for her future. Her self-esteem, happiness and confidence may depend on your commitment.

Parents

- you can show your commitment to your daughter in many different ways, such as: being interested; cherishing, caring and keeping her safe; offering support; making her birthday special, helping her to make sense of her world
- she will feel more secure if she knows you think about her when she's not there: 'Look, I bought your favourite biscuits today!'
- step-parents and partners have to work harder at commitment, particularly when a girl already feels that she has been let down
- behaving consistently, making and keeping to agreements to call, write or visit, are essential ways for non-resident parents to show commitment from a distance

Teachers

- be aware when a pupil is going through a tough time; if she loses a parent or close adult through death, separation or illness, she will need a clear commitment from someone like you – even though she may doubt and test it to the full
- you can show your commitment through patience, tolerance and problem-solving with her when she becomes an adolescent and is not as compliant as she was; there may be simple, practical solutions if you look carefully
- never give up on a student; she may give up on herself, but it is your professional and personal responsibility to continue to offer her hope and belief in herself

Reading on ...

Adams, J *GirlPower: how far does it go?* (Sheffield Centre for HIV and Sexual Health, 1997)

Bryant-Waugh, R and Lask, B *Eating Disorders: a parents' guide* (Penguin, 1999)

Canfield, J and Wells, H *100 Ways to Enhance the Self-concept in the Classroom* (Allyn and Bacon, 1976)

Downes, P and Bennet, C *Help Your Child Through Secondary School* (Hodder and Stoughton, 1997)

Forum on Children and Violence *Towards a Non-violent Society: checkpoints for schools* (1998) [Tel.: 020 7843 6309]

Hartley-Brewer, E *Co-operative Kids* (Hartley-Brewer Parenting Projects, 1996)

Hartley-Brewer, E *Motivating Your Child* (Vermilion, 1998)

Hartley-Brewer, E *Positive Parenting* (Vermilion, 1994)

Hartley-Brewer, E *School Matters ... and so do parents!* (Hartley-Brewer Parenting Projects, 1996)

Hartley-Brewer, E *Self-Esteem for Boys* (Vermilion, 2000)

Katz, A *The Can-Do Girls* (Young Voice, 1997) [Fax: 020 8979 2952]

Katz, A *Leading Lads* (Young Voice, 1998)

Lees, J and Plant, S *The Passport Framework for Personal and Social Development* (Gulbenkian Foundation, 2000)

Lindenfield, G *Confident Children* (Thorsons, 1994)

Noble, C and Bradford, W *Getting It Right for Boys ... and Girls* (Routledge, 2000)

Pipher, M *Reviving Ophelia* (Ballantine, 1995)